MW01156535

PRESENCE 3

GOD, THE COSMOS, PARANORMAL ACTIVITY AND EXOCIVILIZATIONS

THE COSMOBIOPHYSICS 3 THIRDS THEORY

V3.2

Already published by the same author: http://www.denocla.com

Présence 1, Ovnis, Crop Circles et Exocivilisations.

Présence 2, Le langage et le mystère de la planète UMMO révélés.

Présence 3, Dieu, le Cosmos, le paranormal et les Exocivilisations.

Présence 4, Vers un Nouveau Monde…avec les Exocivilisations.

Présence 6, DICTIONNAIRE DENOCLA.

Presencia 1, ovnis, círculos de cultivos y exocivilizaciones.

Presence 2, El lenguaje y el misterio del planeta UMMO revelados.

Presencia 3, Dios, el Cosmos, lo paranormal y las exocivilizaciones.

Presencia 4, Hacia un mundo nuevo ... con exocivilizaciones.

Presence 1, UFOs, Crop Circles and Exocivilisations.

Presence 2, The language and the mystery of the planet UMMO revealed.

Presence 3, God, the Cosmos, the paranormal and the Exocivilisations.

Presence 4, Towards a New World… with Exocivilisations.

UMMO MUSIC Band : http://www.ummomusic.com

UMMO MUSIC, IXINAA

UMMO MUSIC, LIKE 2 OEMMIIs

UMMO MUSIC, BEST OF

Denis Roger
DENOCLA

PRESENCE 3

GOD, THE COSMOS, PARANORMAL ACTIVITY AND EXOCIVILIZATIONS

THE COSMOBIOPHYSICS 3 THIRDS THEORY

ÉDITIONS UMMO WORLD PUBLISHING

This is the third book in the Presence series. It contains new reflections on the contents of the Ummo documents from a research perspective.

Many transcendental mysteries like *"Reality"* or *"God"* are explained into a new rational framework and a new theory is developed.

The REVOLUTIONARY THESES of

THE COSMOBIOPHYSICS 3 THIRDS THEORY

FOR A NEW RATIONALITY

- The COSMOBIOPHYSICS 3/3 theory
- The emergence of life in the cosmos
- The soul and God: rational cosmological concepts
- Telepathy explained
- A rational model for communication with "spirits"
- An elucidation of the "near-death experience"

Translation Kelly Dane, Denis Roger Denocla
Ummodocumentssource:www.ummo-ciencias.org,www.ummo-sciences.org,www.denocla.com and private collections.
Original pictures: special thanks to UMMOAELEWEE.
Digital Illustrations Davy H. —© D. R. Denocla

UMMO WORLD Publishing
8 Esp. de la Manufacture
92136 ISSY LES MOULINEAUX — France

Table of Contents

"Knowledge for whom? Knowledge for what?"

D. R. DENOCLA

ACKNOWLEDGMENTS

I dedicate this book to all OEMMII GAEOAO AIOOYAAO(*) in the WAAM.

I wish to express my gratitude to all those who have kindly helped me to explore new avenues of research.

The books in the Presence series are the result of collaborative effort. All the participants in these works exemplify the willingness to share knowledge in the spirit of humanistic values. The long hours invested by the international volunteer team exchanging information are the mark of their commitment to scientific values and the pursuit of knowledge unhindered by taboos.

I salute the benevolence and altruism of friends and researchers who wish to remain anonymous, as well as Alain Ceria, Anton Parks, Chris Cooper, Kelly Dane, Christopher Blake, Clifford Mahooty, Daniel Verney, Elio Flesia, Gilbert Attard, Jean-Jacques Pastor, Jérémie Filet, Laurent Le Bideau, Manuel Rotaeche Landecho, Marc Pezeril, Marie-Hélène Groussac, Monique Aubergier, Nancy Talbott, Norman Molhan, Thierry Keller, and especially Michel Marcel, the late Gérard Pécoul, and Frédéric Morin of the newspaper *Morphéus*.

D.R. DENOCLA

(*) Feel free to translate this sentence yourself by reading the book PRESENCE 2.

1 2 345678910111213

The COSMOBIOPHYSICS 3 Thirds Theory

Introduction

In *Presence-UFOs, Crop Circles, and Exocivilizations* we looked at the key points of understanding the UFO phenomenon, and the reasons for the terrestrial presence of ET entities in general. We also showed the conditions under which our visitors are traveling. The extraordinary psychological effort to accept these concepts, however, was strongly supported by hard facts and statistics, such as the pictures of the Crop Circles. In *Presence 2—The Language and the Mystery of the UMMO Planet Disclosed*, further evidence is given in the decoding of the UMMO language based on factual material available to everyone. In the first book, we talked about the so-called *Pax Galactica* thesis, the discrete but active and generally peaceful presence of exocivilizations on our soil, the planet UMMO itself, and the opening of a huge field of unknown knowledge, characterized by their mysterious language deciphered in the second book.

This third volume posits similar elements, making the books *Presence 1 and 2* recommended prerequisites to reading *Presence 3.* To fully appreciate the third volume, it is necessary to have the basis of innovative thinking and research perspective from studying the contents of the UMMO documents. The subject matter and this approach is, in itself, already unusual, and requires a real intellectual effort as well as a significant personal psychological investment.

In *Presence 3*, we continue the discussion of earlier works by clarifying concepts and exploring new avenues... In fact, our knowledge of physics, cosmology, and biology should be fully revised. I will explain my reading of the UMMO theory, especially the transcendental cosmological concepts we call *"Reality"* or *"God"*.

We present completely new assumptions, including a possible universal component I call the *"kryptonic constant"* and how it leads to the emergence of Life. We discuss the evolution of the human species according to my reading of the UMMO documents. We deal with some complicated issues currently beyond the understanding of modern science, and reject placing these phenomena in the category of paranormal activity, because that designation fails to explain them.

We explain that these phenomena are quite normal, but understanding them requires a huge paradigm shift to a new rational framework. Thus, we try to understand the possibilities of communication between minds, to what extent reincarnation is possible and within what limits, and we discuss topics related to what is usually called the "soul" or BUAWA. We examine how the stars might affect the human psyche, how humans may eventually develop telepathic communication, and what may be our future evolution. These are some of the many extraordinary and ambitious topics for which this volume proposes to offer explanations.

The author strictly prohibits reference to his research for religious purposes.

NOTA BENE FOR ENGLISH READERS:

Keep in mind that the OOMMO words are written in Spanish phonic, according to the semantic analysis we saw in *PRESENCE 2— The language and the mystery of the UMMO planet disclosed.*

Example, the phonetic word OOMMO will be written here UMMO.

THE COSMOBIOPHYSICS 3 THIRDS THEORY

The COSMOBIOPHYSICS 3/3 Theory is a set of original theses which describe a new rationality. This theory explains all the phenomena remained unexplained by science of the XXI century.

It includes past and present scientific knowledge, rational explanations for the intuitions of Metaphysics, an exogenous vision from the UMMO exocivilization, and numerous developments realized in collaboration with international experts in various fields.

This theory is truly revolutionary because the paradigm which ensues from it cannot be compared to the old world it makes us leave behind ... and this is the difficulty you will face reading it. The theses included in *The COSMOBIOPHYSICS 3/3* Theory are:

- a new paradigm:

  —first third a new cosmological model

  —second third new biology understanding

  —third third new physics

- a new logic
- a new theory of evolution:

  —the emergence of life

  —directed evolution

  —the flow of information from a species

  —the emergence and evolution of man

- rational explanations for:

  —the influence of the planets on the psyche

  —telepathic communication

  —communication with "spirits"

We will discuss the different epistemological contexts over the advance of the presentation. We will compare theses and theories terrestrials with Biocosmophysic thesis presented by our friends UMMO. To facilitate understanding of all, we will illustrate each principle by one or more concrete examples.

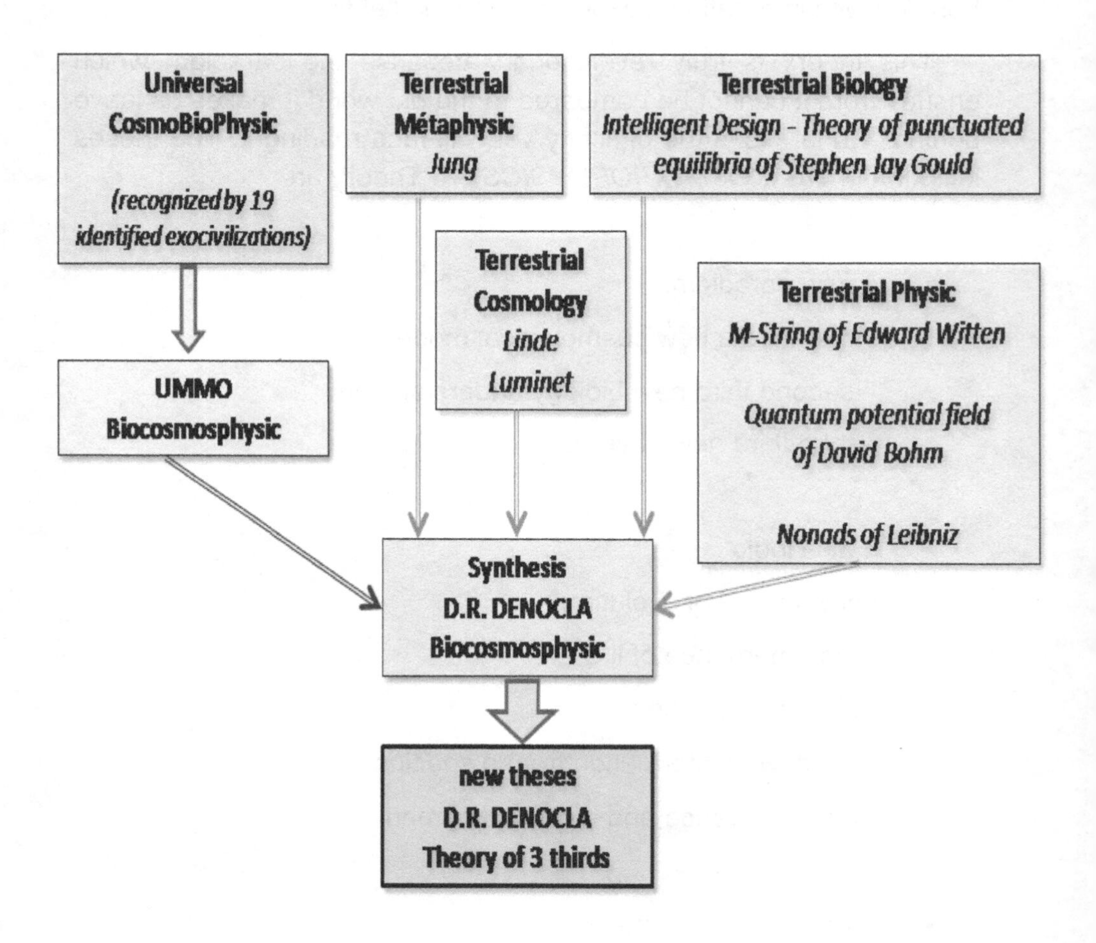

*

1 2 3 4 5 6 7 8 9 10 11 12 13

A NEW COSMOLOGICAL PARADIGM

This presentation of the cosmological framework results from my reading and interpretation of the UMMO documents. We will explain these new concepts with known terrestrial theories and concepts understood largely by theology or cultural traditions for a long time on all continents of the planet... This will be the starting point that will allow us to develop new totally revolutionary theories that unify Physics and Metaphysics.

D28 19/03/1966

You can write any form of written press, magazines, books, provided that it does not distort the reality of our documents. Expressing your views on PHILOSOPHY, RELIGION, HISTORY, SCIENCE and TECHNOLOGY UMMO, you specify which are yours and are not written in our reports.

The contents of this book are very different from the commonly accepted ideas of our time. At best, this is ranked as a fringe or exotic by orthodox cosmologists—we could say *"I smoked the carpet"*—however, the bottom line is that if you want to understand these new concepts, this work is of use to you.

A priori, there are no new ideas from me in this chapter; this is my interpretation of the UMMO cosmology documents whose purpose is a mission of teaching and learning. These explanations are supplemented by important semantic analysis of UMMO words used to help you understand the concepts. This is the fruit of my personal research.

The Multi-Cosmos Model

In the 1920s, the advent of the era of quantum physics, writers and cosmologists began to think about models of the universe composed of multiple cosmos. The idea of parallel worlds existed in fantasy literature before it began to emerge in the conventional scientific framework.

In 1957, Hugh Everett, famous for his many-world interpretation (MWI), presented his vision of the universe using a quantum physics approach. From this point of view the "reality" of the universe splits into many different quantum states, but it is not a multi-cosmos approach.

In 1970, physicists Andrei Sakharov and I.D. Novikov developed the theoretical foundation for a model of the universe consisting of an infinite number of pairs of cosmos/anticosmos. The universe would consist of multiple pairs of sheets or "leaves." As described in the model, at any two places sheets connect in space, they collapse. There is an infinite sequence of layers connected in pairs making up the general structure of the universe. For Sakharov, the pairs of sheets are successive in time, whereas for the UMMO, the "sheets" are concurrent.

In 1990, physicist-cosmologist Andrei Dmitreivich Linde put forth a terrestrial multi-cosmos model comparable to that of the UMMO. He hypothesized that the universe may consist of fractal bubbles in a "foam" structure, each one having its own Big Bang and laws of matter and physical constants; ours being one of those which, by chance, had parameters conducive to the emergence of life. The main difference is that Linde's model proposes the multi-cosmos forms continuously, while in the UMMO model, an almost infinite number of cosmos are initialized simultaneously and then continue on.

The first complete cosmological models consisting of multi-bubbles, called "branes," can be traced back to the work of Lisa Randall and Raman Sundrum in 1999, which in turn was inspired by the work of Arkhani-Hamed, Dimopoulos, and Dvali in 1998. In 1998, physicist Jean-Pierre Petit used the UMMO file to present verbatim their cosmos-anticosmos model. It should be noted that in 2010, Stephen Hawking presented a comprehensive synthesis in The Grand Design, integrating M-theory and the "multiverse" in the sense of the UMMO cosmology.

The multi-cosmos model of the UMMO began to be published in Spain in 1966, and it includes more features than other theories published since then. In 1974, John Schwarz and Joel Scherk developed *"String Theory"* with its connection to quantum gravity, and so did Tamiaki Yoneya in the same year, independently. Yoneya studied the patterns of vibration of the strings describing the bosons, and found that their properties corresponded exactly to those of the gravitons, a theoretical particle that effectuates the gravitational force. Note that the actual UMMO cosmology uses concepts of angular dimensions. UMMO cosmology is quite revolutionary, and as of today, no terrestrial cosmological model uses this type of concept.

In this book, I use the term Universe to indicate the whole Cosmos, to respect the original vocabulary of the UMMO documents. Multi-cosmos was used here, but later the term Multiverse, synonymous with Multi-universe, will be developed.

The cosmological model of the UMMO does not limit itself to the description of a multi-cosmos. It contains other cosmological entities unknown to terrestrial science that we shall present in this work.

The Simplified Cosmological Model

First, here is popularized and very simplified way, the main cosmological entities of the Oommo model are:

- WOA is the generative cosmological entity of "all." The terrestrial metaphysical understanding of this entity is usually called GOD. Effectively, in a simple way, we can say that it creates everything or created everything. This entity exists, but has no mass. We shall see in the following chapters that the ontology of this cosmic entity is most simply termed God. Let us note that WOA is a transcendent cosmological entity.

- WOA concomitantly creates another manifestation that is a little more "material". This is the cosmos called WAAM-UU. This is a super-cosmos, which is the creator and pilot or driver of all other material (physical) cosmoses.

- The WAAM-UU cosmos is thus the origin and generator, the Meta Big Bang, of every cosmos. We only see our own particular Big Bang from the inside of our cosmos. The WAAM-UU is also a "Meta-Meta Brain." Why Meta-Meta? Because this cosmos is a cosmic "Super Brain," which in itself contains a multitude of Meta Brains. Those, in turn, set into motion and guide the systems of evolution that create innumerable life forms on countless planets in cosmoses beyond number. Thus, the cosmos called WAAM-UU is the pilot and wellspring of an almost infinite number of cosmoses that constitute the Multi-cosmos. This concept is not known on Earth and is far from the archaic idea of Gaia.

- The concept of multi-cosmos is close to the model of Andrei Dmitrievich Linde or Lisa Randall and Raman Sundrum, it is called WAAM-WAAM and each cosmos is designated by the word WAAM. We shall see the precise meaning of all these words with semantic analysis. The Multi-cosmos, WAAM-WAAM, is constituted of an almost infinite number of pairs of cosmoses; every cosmos being associated with anticosmos. The UMMO speak of a WAAM and an UWAAM. The Big Bang of every pair gives each cosmos-anticosmos their specific values and parameters of structure, unique to themselves, in particular, the value of the speed of light (C). Every pair of WAAM and UWAAM pursues its life independently of the others, with its own cycle of Big Bang/Big Crunch, according to its unique structural parameters and under the control of the Meta-Meta Brain, pilot of the WAAM-UU.

There are other parts of the cosmological model which we shall present in later chapters. Here is a summary of the first step:

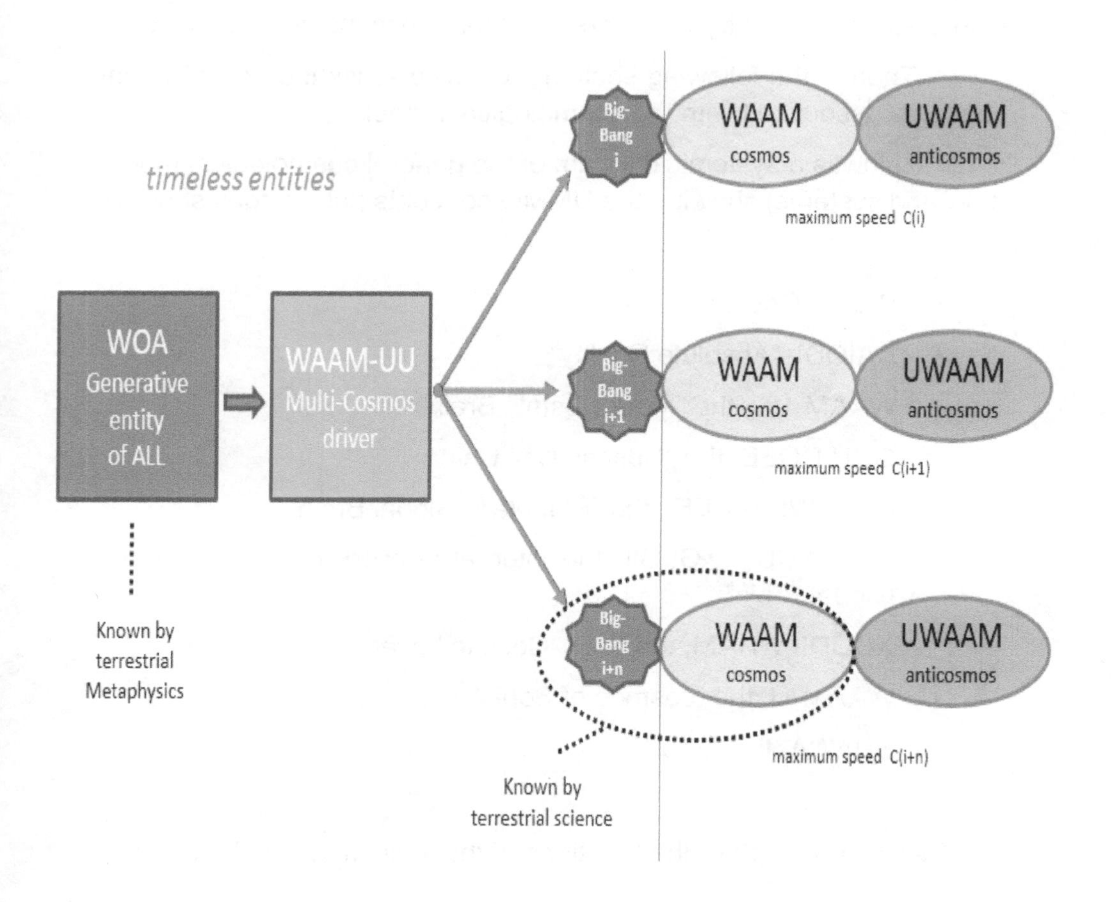
timeless entities
WOA
Generative
entity
of ALL
WAAM-UU
Multi-Cosmos
driver
Big-
Bang
i
WAAM
cosmos
UWAAM
anticosmos
maximum speed C(i)
Big-
Bang
i+1
WAAM
cosmos
UWAAM
anticosmos
maximum speed C(i+1)
Big-
Bang
i+n
WAAM
cosmos
UWAAM
anticosmos
maximum speed C(i+n)
Known by
terrestrial
Metaphysics
Known by
terrestrial science

The General Cosmological Model

We will now go into a more detailed description of the main entities of interest to us, and try to understand their functions in the universe.

Then in the following sections we will discuss the new physical concepts associated with the cosmological model.

Below is a systemic diagram of the general cosmological model (flow and systems) showing the UMMO concepts paired against ours:

- WOA, "God"
- AAIODI, "Absolute Reality"
- WAAM-UU, the "Meta Cosmic Brain"
- GUU DOEE, the "Cosmic Cell Unit"
- BUUAWE BIAEEI, the "Planetary Global Brain"
- GOOINUU UXGIIGII, the internal network of the "Planetary Global Brain"
- XOODII WAAM, the "Inter-Cosmic Layer"
- WAAM-U, the "cosmos of Souls"
- BUAWA, the "Soul"

We will now explore the functions of these cosmological entities...

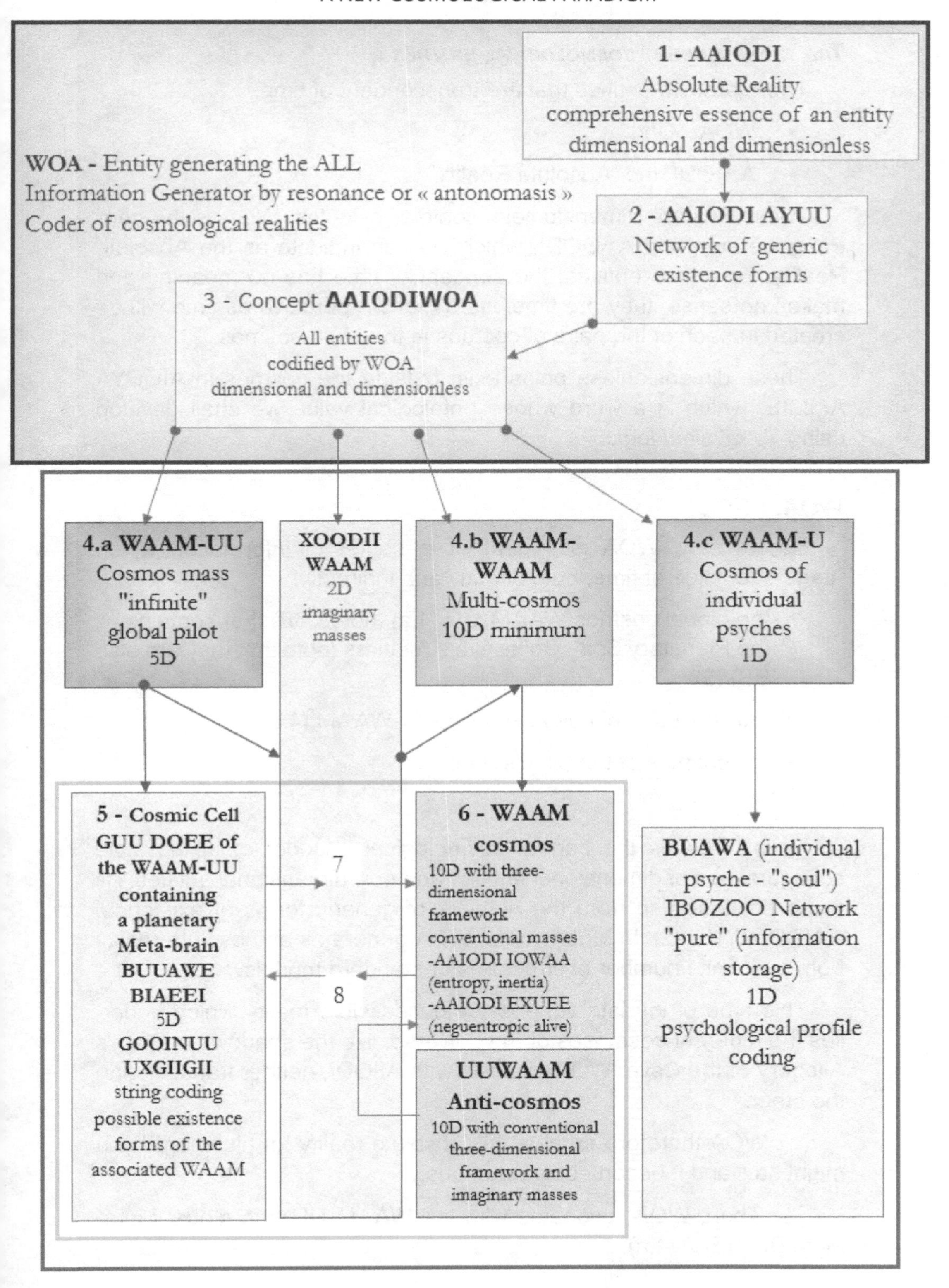

1 - AAIODI
Absolute Reality
comprehensive essence of an entity
dimensional and dimensionless
WOA - Entity generating the ALL
Information Generator by resonance or « antonomasis »
Coder of cosmological realities
2 - AAIODI AYUU
Network of generic
existence forms
3 - Concept AAIODIWOA
All entities
codified by WOA
dimensional and dimensionless
4.a WAAM-UU
Cosmos mass
"infinite"
global pilot
5D
XOODII
WAAM
2D
imaginary
masses
4.b WAAM-
WAAM
Multi-cosmos
10D minimum
4.c WAAM-U
Cosmos of
individual
psyches
1D
5 - Cosmic Cell
GUU DOEE of
the WAAM-UU
containing
a planetary
Meta-brain
BUUAWE
BIAEEI
5D
GOOINUU
UXGIIGII
string coding
possible existence
forms of the
associated WAAM
7
8
6 - WAAM
cosmos
10D with three-
dimensional
framework
conventional masses
-AAIODI IOWAA
(entropy, inertia)
-AAIODI EXUEE
(neguentropic alive)
UUWAAM
Anti-cosmos
10D with conventional
three-dimensional
framework and
imaginary masses
BUAWA (individual
psyche - "soul")
IBOZOO Network
"pure" (information
storage)
1D
psychological profile
coding

THE TRANSCENDENT COSMOLOGICAL ENTITIES

There are two entities that are transcendent of time:

- WOA, "God"
- AAIODI, the "Absolute Reality"

These two are dimensionless centers, or poles. WOA is the central generator, and AAOODI, which we can indicate as the Absolute Reality. For these entities, the concept of time has no meaning and makes no sense; they are timeless. What we perceive as time will be created in each of the pairs of cosmos in the Multi-cosmos.

These dimensionless poles exist outside the cosmos in AIOOYA AMMIE, which is a word whose ontological value we shall develop using *Tetravalent logic*.

WOA, "GOD"

As we saw, WOA is a generative center of information which "begets" outside of time, both directly and indirectly:

- the global cosmos WAAM-UU (4.a) (noted BB) that contains all the Planetary Spirit Collective structures (note that they are also BB) (5)
- multiple pairs of cosmoses WAAM-WAAM (4.b)
- the cosmos of the "Souls" (4.c)

Thus, WOA is the central cosmological encoder of all possible configurations of dimensional entities and non-dimensional entities. All these entities arise from the network of generic forms of existence, AAIODI AYUU (2). In other words, WOA generates an absolute reality from an infinite number of structures or standard models.

This type of infrastructure is an inaccessible reality which underlies the different cosmoses of the universe, like the shadows in Plato's Allegory of the Cave. WOA coexists with AIIODI, neither transcending the other.

WOA therefore creates this absolute reality "in his image" you might say, and it becomes autonomous.

Then, WOA resonates with the WAAM-UU (the global Meta-Meta Brain BB) (4.a)

"... WOA, the center of information is static, while in the multiple universe, the WAAM BB resonates with WOA, and the information is dynamic.

For this, we warn you that the comparison with two violin strings is only didactic and metaphorical, because with them [(WOA and WAAM-UU)-ed.note] resonance effect is manifested by a simultaneous dynamic"

WOA thus models the cosmological entities through the Meta-Meta Brain, WAAM-UU. Standing waves, that is to say constant but distinct phases, correspond to the dimensions created by WOA. In our three-dimensional framework, they form knots and ridges that our senses interpret respectively as "space" and "mass".

D45: "But do not forget that we consider the cosmos as a decadimensional system WOA generates an infinite series of wave trains (the sine function) frequency, amplitude and different phases. SPACE is thus twisted, causing a series of STANDING WAVE NODES which are reflected in the infinity of the WAAM. These standing waves are only the folds of the SPACE-TIME CONTINUUM which we call MASSES (Galaxies, Gas, Animals etc.). This explains the confusion of Earth scientists when they observe the apparent contradiction that an electron is simultaneously CORPUSCLE (mass) and WAVE: this is a naïve confusion."

Therefore, WOA creates standing waves constants, where each wave train establishes and constitutes a dimension. These standing waves underlying the cosmos and form within the non-local universal substratum of the IBOSDSOO, something like strings. We presented this idea in the previous books and will elaborate in this book. Similarly, the idea of a dimension here is very different compared to the current understanding. It will be clarified by the word OAWOO. Let us note that:

"WOA continues to create material inside every Cosmos" (D41-11).

AAIODI, the "Absolute Reality"

The AAIODI (1) is the absolute reality, in the sense of Plato's Allegory of the Cave, and only the shadows of reality appear to us. This absolute reality is autonomous, inaccessible, and an "ACT" of WOA. It is a reality that is behind our warped intellectual perception of absolute reality.

In the existentialist terms of Sartre, the AAIODI is the pure essence of an entity. This may be a dimensional or an adimensional entity.

AAIODI generates all "ideas"—that is to say, the forms of perceived realities—that are compatible with WOA and therefore it contains infinity of ranges or categories, which create a network of generic forms of existence called AAIODI AYUU (2).

THE POTENTIALITY OF CREATING THE ABSOLUTE REALITY

Transcendents cosmological entities create potentialities for the Real Absolute.

It is important to distinguish these transcendent cosmological entities and the Multi-cosmos WAAM-WAAM itself, which itself physically realized those potentialities.

The transcendental cosmological entities are part of the Metaphysics, the Multi-cosmos WAAM-WAAM is part of the world of physics.

The explanation given in the documents of our friends from UMMO:

> WOA (God) creates WAAM-WAAM (the Pluricosmos) at once in its full potentiality (almost infinite).
>
> The EESSEEOEMMII (thinking beings) of WAAM-WAAM materialize some of the potentialities.
>
> Every thinking humanity as a unique part of EESSEEOEMMII changes AIIODII (all potential realities) by interpreting a AIIODI (possible realities).
>
> Each thinking humanity realizes his AIIODI (REALITY) by modulating AIIODII (frame of achievable potentialities). It modifies AIIODII and informs WOA. This information is transmitted and captured through

BOUAWA BIAEII (global Meta brain) associated with each planetary humanity.

Thus the WAAM-WAAM is organized as and when it is generated by WOA. This process is both simultaneous and infinite.

Time takes no part into this process, being only a particular interpretation of each AIIODI.

There is an additional information input carried out on a potentiality experimentally done by a planetary humanity.

For example:

WOA generates the potentiality to enjoy the flavor of food.

WOA generates the potentiality appearance of the fruit "orange".

The two potentialities are realized on planet Earth.

Each terrestrial experiment assessing the flavor of an orange and transmits this information to BOUAWA BIAEII (planetary soul/Meta-brain).

BOUAWA BIAII informs WOA on the overall assessment of the flavor of the fruit "orange" that WOA cannot experiment.

WOA reinforces the potential of the fruit "orange" because its flavor is generally appreciated by OEMMII (human) that can taste it.

The non-infinite character of the WAAM-WAAM potentials based on the only verifiable conjecture by any observer that neither the mathematical zero, nor its inverse (mathematical infinity) exist in the absolute physical setting.

We see that the real absolute interpreted by humanity, AIIODI, is constantly created dynamically, while all his potentialities are almost an infinite tank, but static.

Like many thinkers and researchers have tipped over the centuries, the real world is created based on the idea that we have of it. As strange as it sounds, so we can change—collectively—the evolution of our real, with thinking differently our future... For this it is necessary that the "weight" of our thoughts converge to the desired reality sufficiently... For, example if a large majority of earthlings thought to make a new world governance and ethics, this potentiality would necessarily be realized... (We will discuss this in *PRESENCE 4—Towards a New World ... with Exocivilizations*).

Other Cosmological Entities

The transcendent cosmological entities generate a set of other cosmological entities. Some are known to us, others theorized about, and others are totally unknown to us:

- The WAAM-UU, the Cosmic Meta-Brain
- The GUU DOEE, the Cosmic Cell (Unit)
- The BUUAWE BUAEEI, the Planetary Brain
- The GOOINUU UXGIIGII, the internal network of the Planetary Brain
- The XOODII WAAM, the inter-cosmic communication layer
- The WAAM-U, the cosmos of the Souls
- The BUAWA, the Soul

WAAM-UU, the Cosmic Meta Brain

In a simple and comprehensive way the Cosmic Meta Brain, WAAM-UU, contains all the Planetary Meta Brains, BUUAWE BIAEEI, of all the cosmoses (WAAM) (6).

The WAAM-UU also distorts the other Universes, creating mass singularities, the galaxies, and stars...

"The cosmic plan B.B. [WAAM-UU] contains thousands of millions of B.B. corresponding to so many humanities. This is the B.B. of mankind on Earth, which in connection with your brain processes information received, resulting in the design of things."

That language the UMMO use is based on the functionality of the words. There are words that are functionally homonyms because two objects have the same function ... almost. Thus the WAAM-UU, all the

BUUAWE BIAEEI, and every individual planetary BUUAWE BIAEEI can be noted under the same term of BB.

This may cause some confusion; we will clarify.

"This cosmic plan [WAAM-UU] or B.B. is subdivided into other B.B. or universal psyches, each corresponding to a global humanity (The confusion you might see is just what we call B.B. (BUAUEE BIAEEII) not only the collective Soul UMMO or the Earth, but also the cosmic plan (that is to say the multi-verse) that contains all the B.B. of different social networks that populate our tetradimensional Universe."

In detail, the WAAM-UU contains an almost infinite number of cosmic GUU DOEE cells. Each cell contains a BUUAWE BIAEEI (global network/brain)

The Cosmic Cell GUU DOEE

The cosmic unit, or cell called GUU DOEE contains a planetary pilot/brain, BUUAWE BIAEEI. The term GUU DOEE means:

GUU = the structure has a dynamic dependence

= a hermetic structure

DOEE = the shape of the entity has a model

= a modeled shape or form

The translation of GUU DOEE: *"the hermetic structure has a modeled form"*

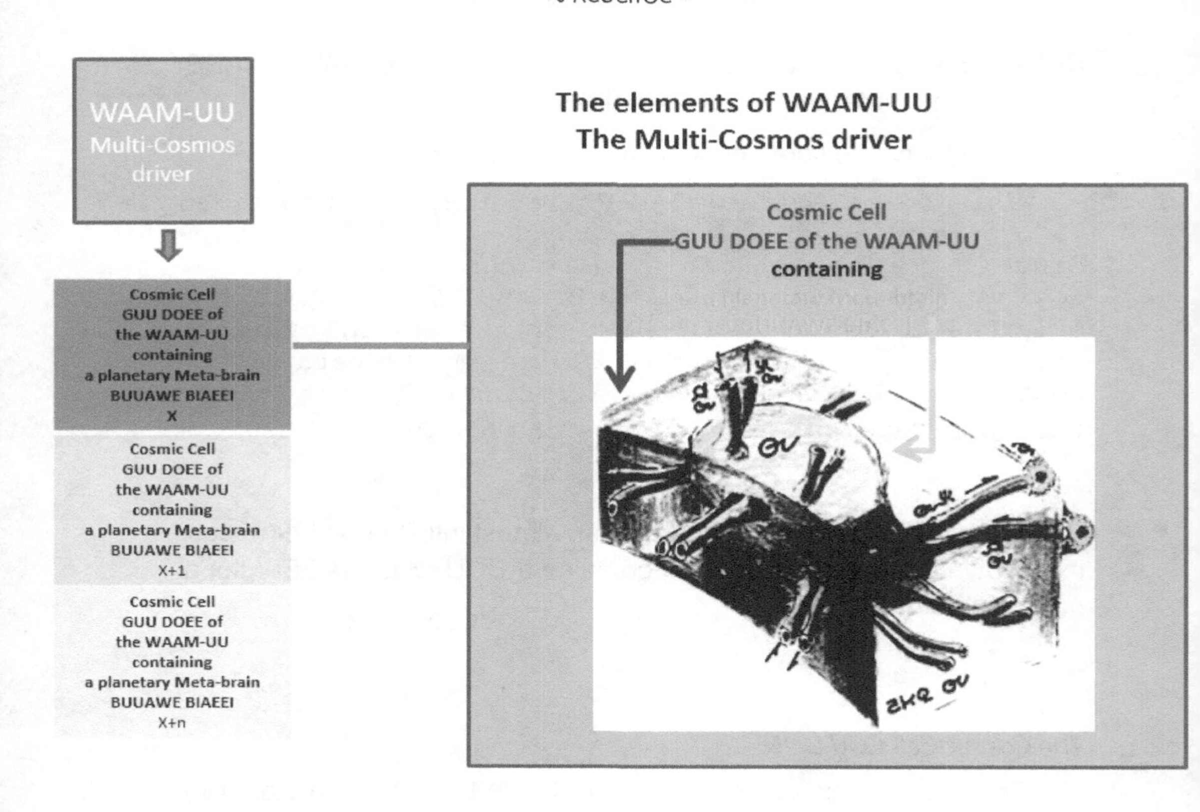

D357: The B.B. [WAAM-UU] is constituted by GUU DOEE (outlines or cell units). An image to help you understand us would be, "the galaxies of our Universe" except in B.B. there are no nebular configurations of dust and suns, but pregnant, five-part dimensions.

Said differently: this cosmic plan [WAAM-UU] or B.B. is subdivided into other B.B. or universal psyches [in the GUU DOEE]

The Planetary Brain BUUAWE BIAEEI

The first formalized concepts approaching the concept of "global brain" appear from the early twentieth century. Swiss psychiatrist Carl Jung understands that the human psyche is not limited to the skull. It is necessarily a provision of external information. This information seems to be the major profiles of cultural values, "universal" knowledge. This suggests to Jung that there is an informative exogenous structure to humans, which provides the archetypes at a deeper level, a collective unconscious shared by the entire human species. This idea will unfortunately not be developed by other scientific disciplines, and in this case will be totally ignored by biologists...

In 1936, Vladimir Vernadsky developed the concept of noosphere, or sphere of thought, in a model of the inheritance of the Earth development phases, and offers five different layers of interaction:

- the lithosphere, rock core and water;
- the biosphere consists of life;
- the atmosphere, gaseous envelope constituting the air;
- the technosphere resulting from human activity;
- the noosphere, or sphere of thought.

Just as the emergence of life fundamentally transformed the geosphere, the emergence of human cognition fundamentally transforms the biosphere, the Vernadsky's noosphere emerges when humanity by controlling nuclear processes, begins to create resources through the transmutation of the elements...

At the same time, in 1932, in "Christology and evolution", Pierre Teilhard de Chardin spoke of "sphere of human thought". For him, the "human phenomenon" should be thought of as being—at some point—a stage of evolution that led to the "sphere of human thought" which prepares the advent of the figure called the "Cosmic Christ ". The "point Ω or Omega point" represents the convergence pole of evolution. Omega forms somehow attraction at stake in the individual as much as collective level.

In 1970, James Lovelock developed the theory that the Earth would be "a dynamic physiological system that includes the biosphere and keeps our planet for over three billion years, in harmony with life." All living things on Earth would be like a vast superorganism called "Gaia" realizing self-regulation of its components to support life.

For several decades, the properties of psychometrics or retrocognition are used with some success by archaeologists. These properties or abilities of individuals to "see" in the past mediums such as Gérard Croiset, Eillen Garret, Hella Hammid, George Mc Mullen, etc. made extremely precise descriptions of ancient sites and allowed numerous archaeological finds. A multitude of cases are reported by Stephan A. Schwartz in The Secret Vaults of Time (New York, Grossel & Bunlap 1978)

Most theologians from all religions, in all ages, pertinently observe phenomena NDE—Near Death experience—reported by the dying "resurrected". As we will discuss in detail in Chapter NDE, this phenomenon proves the existence of an afterlife. Logically, these theologians conclude the existence of God. Obviously it is not the transcendent entity which is designated herein as WOA. It is simply this cosmological object, although physically located in another cosmos, "global planetary brain" BUUAWE BIAEEI, which is unknown to earthlings.

Karl Pribram, neurophysiologist of Stanford University influenced the physicist David Bohm, who will sense the existence of collective unconscious common to all of humanity. "In his deepness, humanity is one and the same psyche. ". Nevertheless, too rooted in his convictions and without cosmological vision, he does not perceive the possibility that it may be a physical object and cosmological located in another cosmos...

The "global planetary brain" BUUAWE BIAEEI contains all the information that all living beings have transmitted to him since the dawn of time, specifically since our "global planetary brain" BUUAWE BIAEEI is connected to the Earth. And therefore particularly archetype coding structures (GOOINUU UXGIIGII) of all living beings on the planet. It pilots the coordinated co-evolution of living beings. If necessary, it could decide to eliminate a terrestrial or an extraterrestrial pest in vital ecosystems functioning of the planet. For example, even Human if it becomes a mortal danger to all other living species, then BB eliminate this pest...

The way a planet guides or steers alive beings through the BUUAWE BIAEEI is a cybernetic loop that can be simply summarized as:

a) sending structural information for semi-autonomous entities to use

b) feedback from the entities

This cybernetic loop has a permanent dynamic.

D731:

"The alive beings capture the structure of the Universe through their transducers, the receptive neurosensors of models of information (their sense organs).

This information is sent to the B.B. and integrated and processed in the WAAM-UU.

Which, in turn, engenders models of action on the WAAM WAAM. This closes the cybernetic loop."

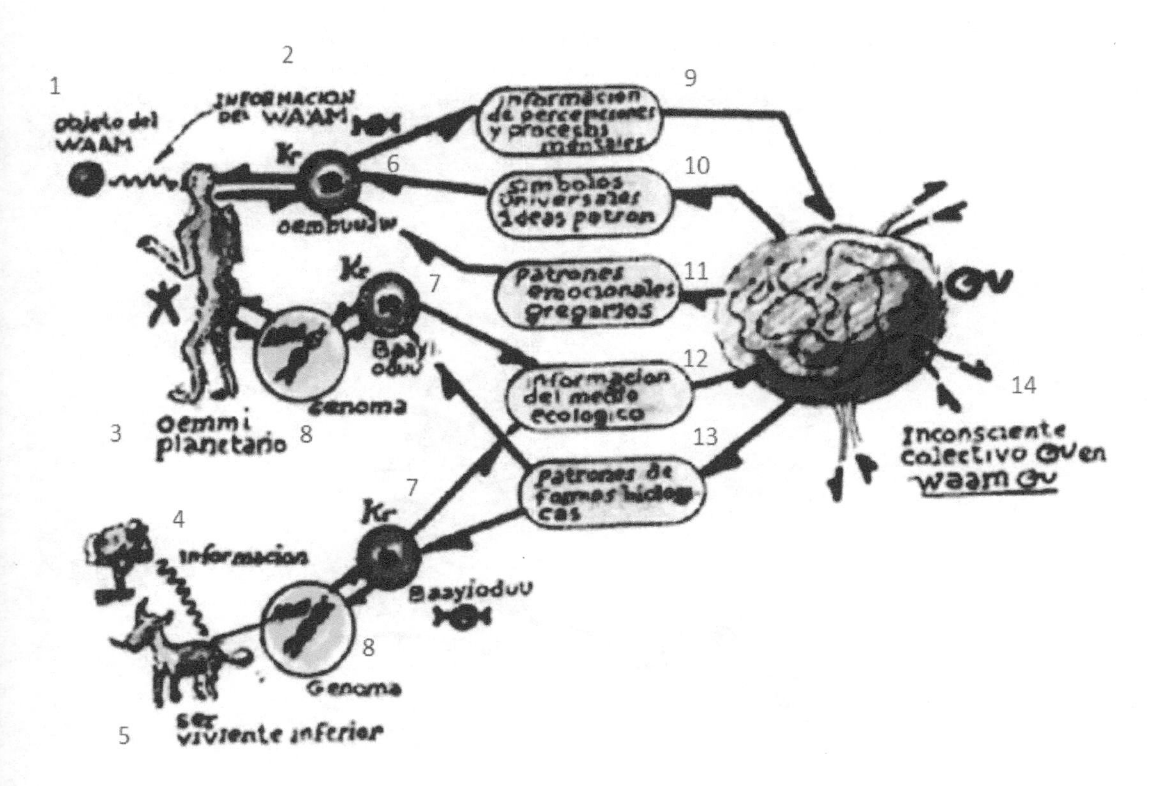

1. objects of the WAAM
2. information of the WAAM
3. people of the planet
4. information
5. lower life forms
6. OEAMBUUAW (cerebral-cosmic emitter/receiver)
7. BAAYIODUU (genomic-cosmic emitter/receiver)
8. genome
9. information of perceptions and mental processes
10. universal symbols, idea-patterns
11. group emotional patterns
12. information of ecology
13. patterns of forms
14. the collective unconscious BB in the WAAM-UU

"The WAAM is a pentadimensional continuum with mass singularities (in the form of filament nodes), divided into 'cells' or 'environments' separated from one another."

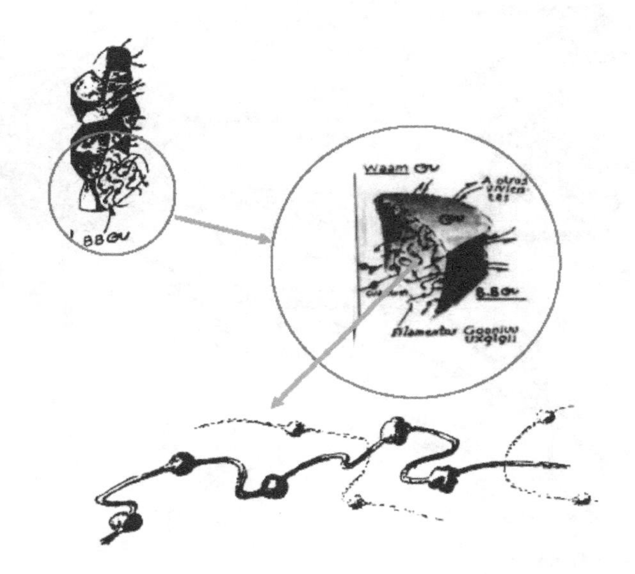

The pentadimensional structures of the GUU DOEE contain a planetary brain pilot BUUAWE BIAEE, which itself contains three dimensional filaments space and mass (+M and—M), and the two-dimensional GOOINUU UXGIIGII (5) where information circulates.

Each planetary Meta-brain BUUAWE BIAEEI (5) is connected to a star and contains:

- the guiding biological pattern of alive beings
- universal generic "ideas»
- collective sentiments
- group behavioral profiles
- moral ideas, the moral guides of higher life forms (OEMMII)

Each cell receives the name BUUAUE BIAEI (BB) "Collective Soul or Spirit." There are as many BB as AUUBAAYII (networks of alive beings on a planet) throughout the WAAM-WAAM. There is a one to one correspondence between each set of alive beings on a cold celestial body and the respective BB.

The information transmitted to entropic beings, inert AAIODII IOWAA

The planetary BUUAWE BIAEEI transmit information to entropic entities—inert AAIODII IOWAA—via an inter-cosmic layer, the XOODII WAAM (6), containing imaginary masses (+Mi and—Mi) which are responsible for the many gravitational interactions affecting pairs of cosmoses, including ours, evidently. Also, the BB allows the extraordinary effect of exchange between cosmoses, called LEEIIYO WAAM. For example, thanks to the decoding of UMMO words in Presence 2, we were able to understand what we called the "trampoline effect" used by most interstellar craft.

Information sent to negentropic beings, living AAIODII EXUEE

The information sent to negentropic beings—living AAIODII EXUEE—is made through two channels:

- for all alive beings, by an intracellular communication channel, a BAAYIODUU (7) (genetic-cosmic transmitter/receiver)
- for human beings, by a cerebral communication channel, an OEMBUAWE (cerebro-cosmic transmitter/receiver)

The intracellular communication channel associated with the genomic system, BAAYIODUU (7), establishes the generative factor of Life, the BAAYIODIXAA. *(See the Emergence of Life hypothesis).*

"Each B.B. sends its biological patterns to guide sentient beings' evolution (ORTHOGENESIS) on each cold planet".

The negentropic beings, living AAIODII, send information back to their associated planetary BUUAWE BIAEEI (8) via the intracellular communication channel, the BAAYIODUU.

The human brain's communication channel, the OEMBUAW, receives and returns the information to its planetary BUUAWE BIAEEI.

"Each B.B. also ships its universal ideas, collective feelings, social inductions, moral patterns, etc. to all OEMMII [humans]"

This information is transmitted directly through an imaginary mass, via a cerebral channel, the OEMBUAW (6).

—Is there any way to recognize when a mental process, a thought, comes from the BB?

U. —NO. The information reaches the deeper layers of the brain, and so it is very difficult to distinguish this information from something learned in childhood.

In their thinking, the higher orders of life distort the absolute reality. The information is transmitted directly through the cerebral channel OEMBUAW.

And ... "if we accept the definition of WAAM-WAAM in the strict sense, there must be as many categories of thinking beings capable of distorting the AAIODI as there are WAAM [cosmoses]."

The Aura and OEMBUAW

Journalist Michael Talbot cites in his book "*The Universe is a hologram,*" the work of Valerie Hunt Professor of Physiological Sciences at the University of California, Los Angeles. She discovered that the energy field responded more quickly to stimuli than the brain. The parallel connection of an electromyograph and an electroencephalograph on tested subjects showed her a marked delay in the registration of the second when happen sudden sound and light variations. Valerie Hunt thought that the mind was not in the brain but in the aura.

Our hypothesis is different and more complex. We believe that the spread of information reaching the OEMBUAW is through neural circuits and frequency emissions. As we shall see later, probably via the relay compound, a bio-tuner, GeSi2C3H3 which acts as an interface "tuner" multifrequency with pairs of Krypton. In that, the wave propagation is faster than propagation into the bioelectrical circuits of neurons.

Thus, the human bio-field, the Aura, receives the information from the OEMBUAW before information treatments centers in the brain

like the neo-frontal cortex, for example...

The frequencies of the human bio-field, the Aura, are chronologically the first informed of the details of the planetary BB and BUAWA flow. It is almost a direct connection with these cosmological objects. This will be important in the process of death of the human, the Aura can remain connected with the Meta-brain BB and/or BUAWA Soul.

To summarize, according to the systemic plan of the general cosmological model, we could say that the planetary BUUAWE BIAEEI are systems of directing cold planets, where alive beings are effectively the sensors of the operating system. They send information to the piloting system. In accordance with the Shannon-Hartley theorem, the complexity of the control is higher than that of the operating system.

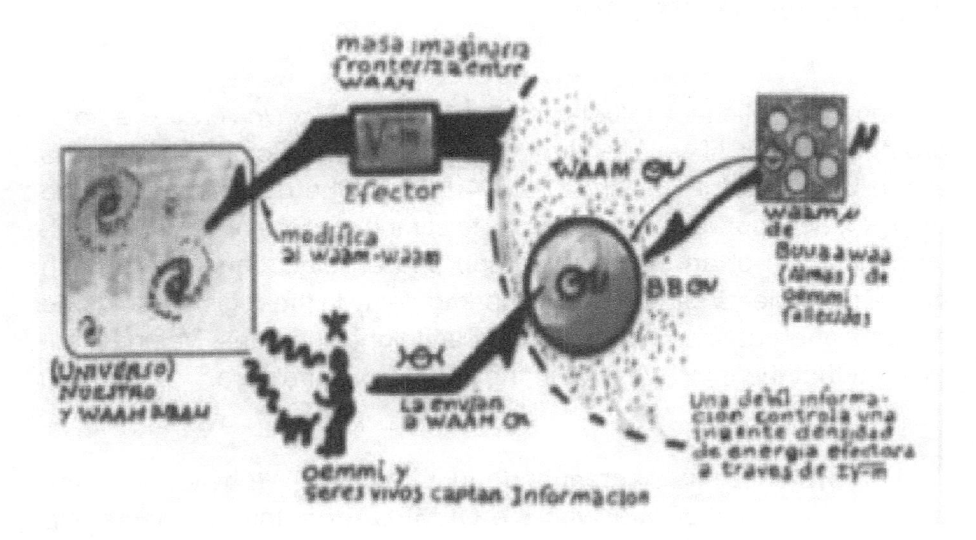

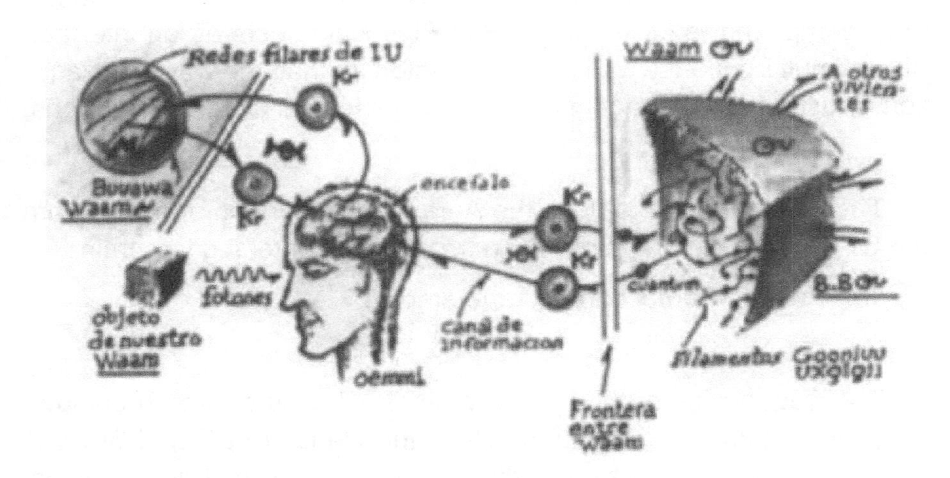

The inter-cosmic communication layer XOODII WAAM

Simply put, the XOODI WAAM is the inter-cosmic conveyor of messages. This is the communication channel: the link to connect all the different cosmological entities.

The XOODII is a space 2 angular dimensions OAWOO. A relay layer on which normal masses, UFOs for example, can bounce effect by generating a LEIYO we called “trampoline effect” in “PRESENCE1—UFOs, Crop Circles and Exocivilizations” and which is interpreted as an anti-gravitational effect on the Earth paradigm of XX—XXI century. But there is indeed something else, unknown, to contemporary science.

This XOODII transmits multiple effects called "LEIYO" between different cosmological entities of distinct nature. It transmits gravitational effects of the masses of the torque anticosmos-cosmos. Explaining the "dark matter" and "dark energy" variations in the cosmic expansion.

—Why don't beings made of imaginary mass exist?

U– "Because for them, time would pass in the opposite direction. If there were beings of imaginary mass, their memory would be of the future. It would be absurd; time would flow backwards for them, so the present would manufacture the past."

In the cosmos, as in the anti-cosmos, time is the result of the axes of the three-dimensional spatial framework. Time is related to the speed of light (C) in this setting.

In the XOODII, the speed of communication is between C and infinity. In other words, if a hypothetical object in the XOODI moved faster than the limit of our universe, the speed of light in our cosmos, then that object would move faster than time runs in our cosmos. The object would thus see time going backwards...

WAAM-U, the cosmos of the "Souls"

The WAAM-U is the cosmos of individual psyches. In terrestrial metaphysical terms, it would be called the cosmos of souls. This is something separate from the Collective Spirit, but comparable to the planetary Meta-brain BUUAWE BIAEEI.

BUAWA, the "Soul"

The cosmological entity BUAWA is called the soul in terrestrial metaphysics. It is a complex entity which is formed at conception.

"In principle, we know that all living organisms, an alga, a bacterium, or a giraffe of the Earth, when it generates, has a BUUAUUA (individual soul) (B) in the "distant cosmos". This soul is barren. Its network of IBOZSOO UHUU is not able to codify any information because there is no link that attaches to the body that was born on a cold star. Only OEMMII (human) synthesizes at the moment of chromosomal fusion a set of Krypton atoms that, through a process that we call effect or border membrane LEIYO, allows communication between two such distinct Cosmos. «

"When a human is 'born', that is to say, not at the moment of Parturition [delivery], but when both male and female gametes deliver their genetic load, an enormous cell of IBOZSOO UHUU develops (actually a complex network of these particles are formed by a large chain of angular relationships). These large chains in turn form a broad substrate or matrix where all the information of our lives is encoded in a sector of the network, while the remainder codifies a whole set of instructions, or program, that conforms to each tetradimensional OEMII."

"The BUUAWEA has no memory, nor is it capable of feeling or perceiving. It cannot, for example, move or feel pleasure or pain."

"[it does, however,] ... generate ideas, and is able to comprehend the messages brought to it by the OEMBUAW, whereby it is also able to act and control the OEMII (body)."

"The ideas generated, the knowledge acquired, the control of the body DOES NOT MANIFEST IN A SEQUENTIAL MANNER OR IN A CONTINUOUS FLOW OF TIME.

IT MAY MODIFY ONCE AND FOR ALL THE FORM OF THE OEMBUUAOEMII (HUMAN FORM IN SPACE-TIME)."

" ... the BUUAWEA has the ability to shape the conduct of the body throughout time, once and for all."

"The soul does not think. THE BRAIN THINKS. The soul stores data, and by the interaction of sequences of I.U. and the cortical neuronal networks, governs the spatio-temporal behavior of the human organism (WILL)."

To summarize, BUAWA, the "Soul", consists of two zones.

The first zone is the sector of a network of "pure" IBOZOO UU which has the function of storing information. This area is "formed by large chains of angular relationships. These large chains in turn form a broad substrate or matrix where all the information of our lives is encoded."

The second zone of the network of pure IBOZOO UU is "mental conformation" which is performed once and for all in its completeness. This zone codifies a whole program of instructions that shape each OEMII (only mankind takes on the neuronal dimension OEMII + BUAWA = OEMMII).

We can deepen our understanding of this entity by the semantic analysis of the word BUAWA *(see details in Presence 2).*

For BUAWA, we have this transcription:

- The interconnection depends on the movement which generates a displacement
- The interconnection depends on the movement of an electron in the atom chain of Krypton, OEMBUAW, which generates a displacement (an act of will).

More simply:

- The interconnection depends on the movement generated by an action
- Interconnection generator

In conclusion, BUAWA is the generator of the human will, the conduct of the human body. It contains two parts:

- BUUAWA IMMI is the global awareness of all the events and experience of the lives of OEMMII. The first zone is the area of pure IBOZOO UU which stores information, and records all the information of our life.

- The ESEE OA is the consciousness of the present moment. The second zone of the network of pure IBOZOO UU is a "zone of psychic conformation" which is done once, comprehensively and permanently, but which manages awareness of the present moment. This zone "codifies a whole program of instructions that shape each OEMII".

Synthetic Scheme

At this point in your reading, it must be becoming clear that the mysterious image chosen for the cover is actually a depiction of a planetary Meta-Brain BB, a human being in a 4D Space-time, and his/her "Soul" ...

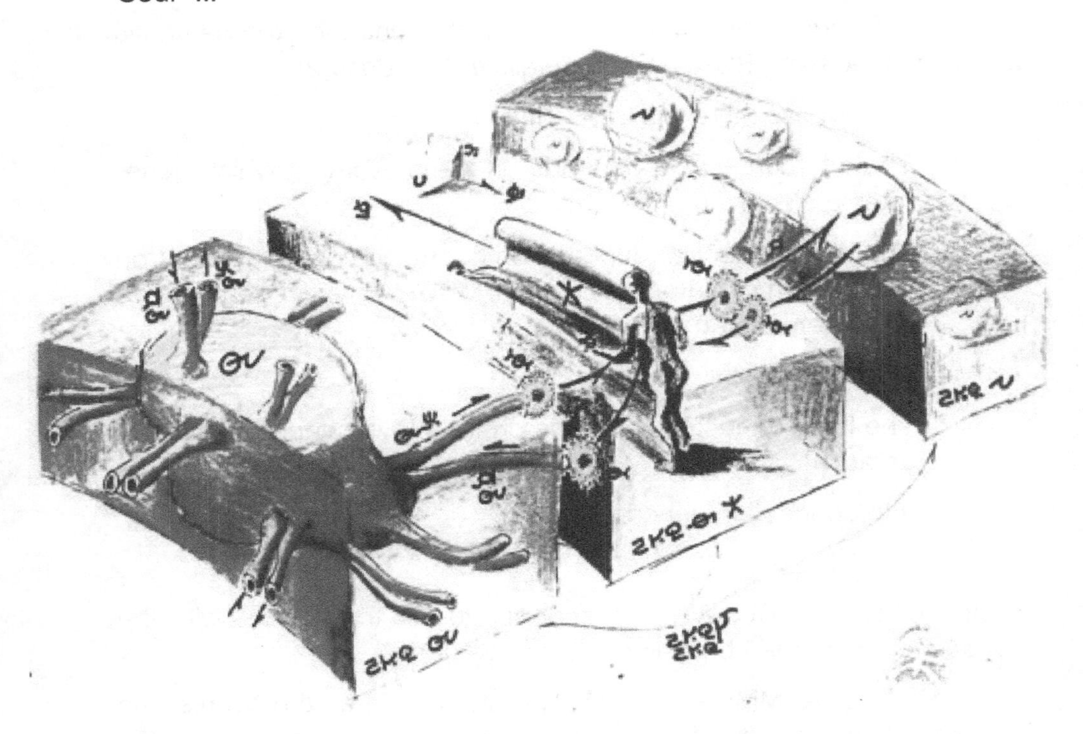

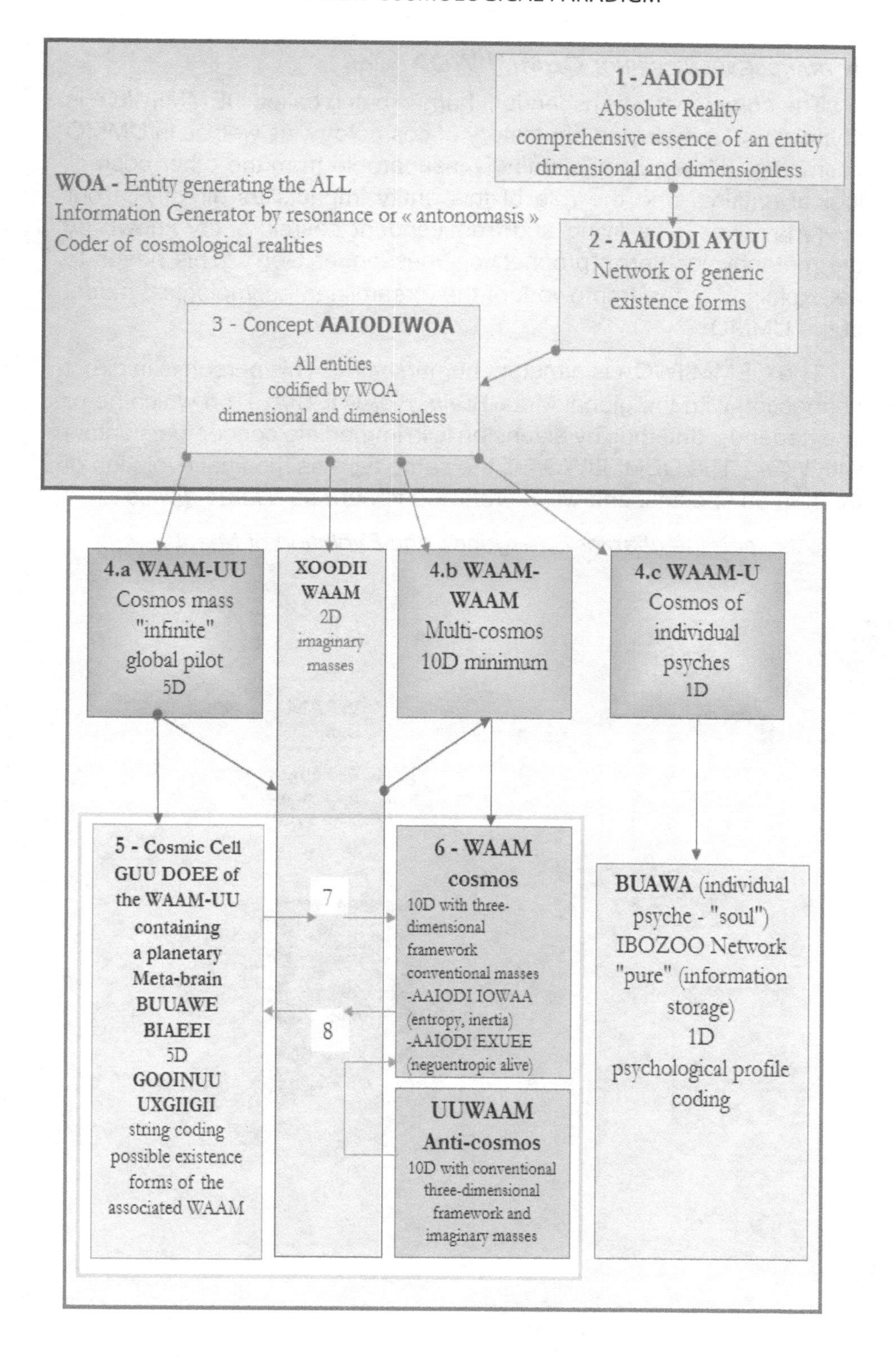
1 - AAIODI
Absolute Reality
comprehensive essence of an entity
dimensional and dimensionless
WOA - Entity generating the ALL
Information Generator by resonance or « antonomasis »
Coder of cosmological realities
2 - AAIODI AYUU
Network of generic
existence forms
3 - Concept AAIODIWOA
All entities
codified by WOA
dimensional and dimensionless
4.a WAAM-UU
Cosmos mass
"infinite"
global pilot
5D
XOODII
WAAM
2D
imaginary
masses
4.b WAAM-
WAAM
Multi-cosmos
10D minimum
4.c WAAM-U
Cosmos of
individual
psyches
1D
5 - Cosmic Cell
GUU DOEE of
the WAAM-UU
containing
a planetary
Meta-brain
BUUAWE
BIAEEI
5D
GOOINUU
UXGIIGII
string coding
possible existence
forms of the
associated WAAM
7
8
6 - WAAM
cosmos
10D with three-
dimensional
framework
conventional masses
-AAIODI IOWAA
(entropy, inertia)
-AAIODI EXUEE
(neguentropic alive)
UUWAAM
Anti-cosmos
10D with conventional
three-dimensional
framework and
imaginary masses
BUAWA (individual
psyche - "soul")
IBOZOO Network
"pure" (information
storage)
1D
psychological profile
coding

A transcendent being: OEMMIIWOA

The concept of a transcendent human being called OEMMIIWOA is atypical and marginal in our history of cosmology, as well as in UMMO cosmology. However, this entity is inseparable from the other cosmological entities, and the role of this entity impacts us directly in our everyday lives. The living and transcendent cosmic entity known by the metaphysical term "prophet" or "messenger God". This needs to be explained in the framework of the streamlined cosmological model of the UMMO.

The OEMMIIWOA is a mutant human being. This person is in direct connection with the global Meta-Brain, BUAWE BIAEI, on which he or she depends, and thus by extension is in immediate conceptual contact with WOA. The OEMMIIWOA is the same type as the final evolution of the human species, and will be referred to here as "*Homo divinis*".

(Also see the chapter Emergence and Evolution of Man.)

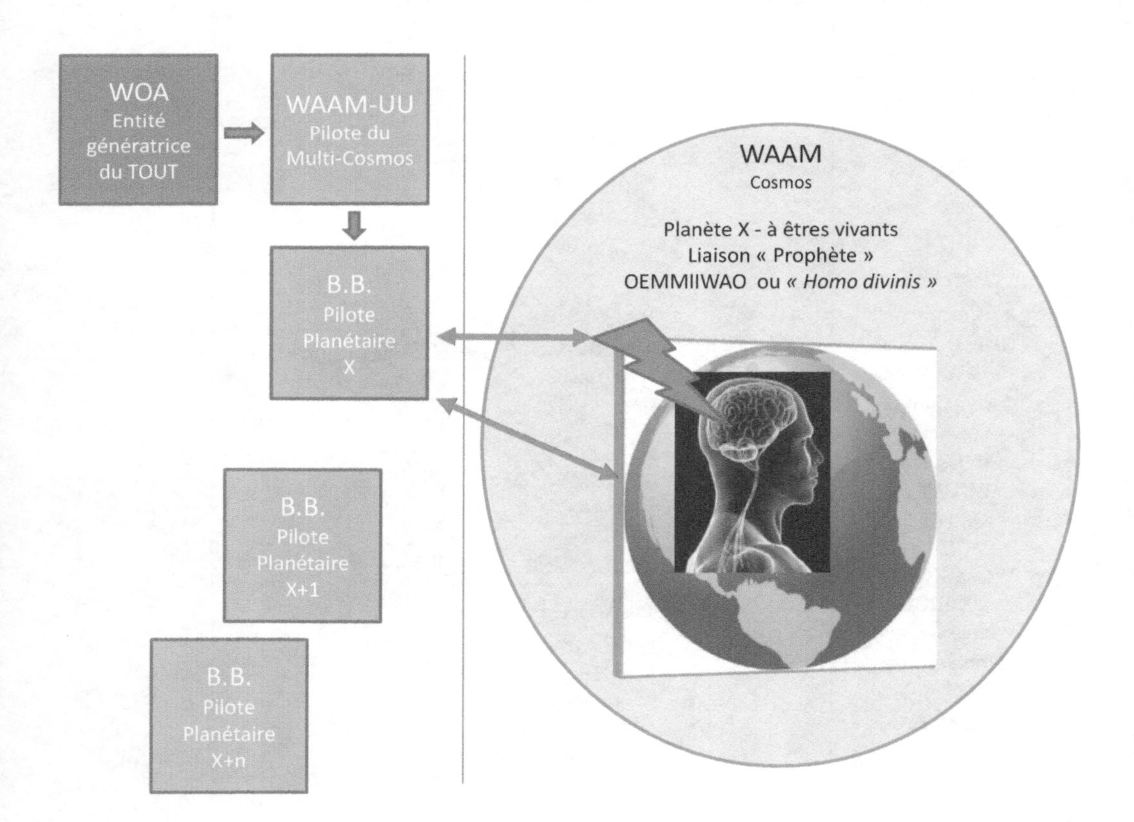

The semantic analysis of the phonetic term "oemiwoa" sheds light on the function of Homo divinis, who is seen as a "prophet" in a world of Homo sapiens.

The concept of OEMMIWOA is not exactly the general concept of a prophet; that term would be UMMOWOA for the prophet of UMMO, and OAYAGAAWOA for the prophet of the Earth. In this case, according to the UMMO documents, the only known mutant individual of this type on Earth was Jesus Christ.

The word OEMMIIWOA generally refers to "a human brain with a singular connection with the divine". A set of cerebral mutations gives the person direct access to a new information channel with the cosmological global planetary pilot, B.B.

The translation of OEMMIIWOA in context:

- (body bounded by the physical envelope) who has access to (the generator of displaced entities)
- God is the human
- Human connected with God

In other words, the final human species: "*Homo divinis*"

D792: Cosmo-biological STRUCTURE: In an OEMIIWOA, the conventional network of BAAYIODUU arises, built by Krypton atoms, which put his brain in connection with its BUAAWAA and the BUAWWEE BIAEII BAAYIODUU. However, in addition, there is a new network with a number, unknown to us, of Kr (Krypton) atoms that establishes an informative connection with the cosmic pole of WOA information.

FUNCTION OF THE OEMIIWOA: In the brain of the encephalon having undergone such a mutation, is treated at an unconscious level, the UAA of WOA. The AYUUEAODII or emergence function of this nervous system, manifests itself with a great intensity of the wonderful plan of Cosmological Nature. This encephalon has a mission to inject into the social network these negentropic laws, into encephalon endowed with free will, capable of accepting them or pushing them away. The information concerning these laws (morality, you would say yourselves) channels from this brain into the global body of data which forms the cultural heritage of this society.

The OEMMIIWOA is converted thus into a direct receiver of certain rules of order contained in WOA, and not only through the B.B. as for the rest of human beings.

The chromosome structure of the OEMMIIWOA prevents procreation with an OEMMII because both are of different species. [...] this OEMMIIWOA ignores that his body is different from other intelligent humans, given that the anatomy is very similar".

THE COSMO-PHYSICS

We explored a comprehensive cosmological vision in view of the very big, and it must be complemented by an overview of physics in the context of the infinitely small.

We will next explore the objects of physics related to this cosmology.

THE UNIVERSAL SUBSTRATUM

D117: "We call IBOOZOO UU, entities whose suite is linked together by various angular rotations. They may have energy characteristics of mass or space, depending on rotations corresponding to elements of that sequence."

The universe, composed of almost infinity of cosmos, would consist of a substrate of multidimensional elements, quite similar to infinitesimal strings presented in the M-Theory String of Edward Witten. These multidimensional elements are non-local, that is to say, they are subquantum, without the concept of time, without a notion of space, without a notion of force or energy. Because time, space, energy or forces, are emerging from the non-local subquantum substrate...

For a long time men have suspected the existence of this universal substrate. But they sensed it, only with the view of a lonely cosmos.

This universal substrate is a kind of large virtual mesh that was interpreted by the great Indian thinkers as an illusion. The world was an illusion emerging from the Vedic Maya.

As stated in the Svetasvara Upanishad: "You should know that Nature is Maya, illusion, that Brahman is the illusionist and that this world is peopled with beings who take part in his presence."

Almost all cultures have sensed this, it would take too long to make an exhaustive history, but closer to us in time, in the seventeenth century Leibniz, following the Pythagoreans, saw the origin of the cosmos made of fundamental entities, called by him "monads", each of which was as a reflection of everything and could only be defined by its relationship with the other monads. For Leibniz, every being is either a monad or a compound of monads.

In nature, the monads are simple substances endowing "appetition and perception". As to their structure, they are per se units, analyzable into an active principle called "soul", "substantial form" or "entelechy" and a passive principle, called "mass" or "raw material". As for their expression, monads are each living mirror, representative of the universe, according to their views.

As to their hierarchy, the monads have degrees of perfection: the lowest degree, simple monads or "naked" are characterized by unconscious perceptions. They contain all the information on the state of all the others but have neither conscience nor memory... This approach led Leibniz to invent the integral calculus.

A few centuries later, thanks to mathematical discoveries Fourier in 1947 the physicist Dennis Gabor invented the principle of holography. This is a recording method of phases and the amplitude of the diffracted wave by an object. Diffraction is the result of interference of the waves broadcast by each item and includes a resolution limit, the minimum distance or angle that must elapse between two contiguous points so that they are correctly discerned by a system of measurement or observation.

This recording method makes it possible to subsequently render a three-dimensional image of the object, the hologram. To record it, you must code amplitude and phases of the light coming from the object considered. For this purpose, two coherent beams caused to interfere on a photographic plate. The first beam, called the "reference wavelength", is sent directly to the plate. The second, called "object wave" is sent to the object to be photographed, which diffuses the light in the direction of the photographic plate. The interference figure thus formed contains all the information about the amplitude and phase of the object wave, that is to say the shape and position of the object in space.

This principle inspired the physicist David Bohm's holographic model subquantum in 1952, which will partly be the earthly forerunner of the IBODSOO model.

Previously, the great Danish physicist Niels Bohr noted strange quantum phenomena interconnection between particles that will be called later "quantum entanglement phenomena". Niels Bohr thought that if the elementary particles did not exist before being observed, conceive them as independent "objects" had no meaning. Talk about their properties and characteristics as existing objects observation was losing its meaning. It could not be more disconcerting. For Albert Einstein, Boris Podolsky and Nathan Rosen, was unacceptable and they even published a famous article "Can we take for complete description of reality by quantum physics?" explaining that there could be no interconnection between particles faster than light, which is known as the Einstein-Podolsky-Rosen paradox.

Taking into consideration the observations of Niels Bohr and Albert Einstein, the plasma physicist David Bohm considered that particles like electrons have a tangible existence in the absence of any observer. And his knowledge of plasmas let him also assume a buried reality underlying, a subquantum map unexplored by science. He named this new field "quantum potential" and attributed to him, as to the gravity, the theoretical property of being everywhere in space. This enabled him to understand that the plasma electrons can have the globalities behavior interconnected in the sense of Bohr. It notes that the plasma electrons are grouped, through the quantum potential, it's the whole system that performs a coordinated move closer to the choreography than random swirls of a crowd. This activity is closer to the operation of the various parts of a living organism as the joining of parts of a machine. The interpretation given by David Bohm of quantum physics suggests that subquantum plan in the field of the "quantum potential", any location ceases to exist. Each point of space is consubstantial to all others and talk about anything as distinct from what everything becomes absurd.

This is what is called "non-locality". The non-local "quantum potential" aspect of David Bohm allows him to explain the connection between twin particles without violation of the relativistic ban imposed onto any transfer to a speed greater than that of light...

Experiments bringing conclusive evidence of what is now called quantum entanglement, will come in 1982 by the physicist Alain Aspect and his team at the Institut d'Optique d'Orsay-Paris XI. They demonstrated that atomic particles of our physical world are entangled in a non-local way. So there existed a non-local subquantum substrate...

Joining the concept of the Maya, David Bohm developed a theory of the holographic universe, thinking that the reality of our world is the expression of a hologram, that is to say, a non-local holographic substrate. An electron is not an "elementary particle", just a name for a certain aspect of the dynamics of a hologram substrate. According to him, an order "implied", governed by holographic principles also explain the non-local appearance of the real at local subquantum level. And if the structure of the universe was that of a hologram, why be surprising whether it has non-local properties?

By logical abuse David Bohm thinks that eventually every point of this substrate of the universe could contain the entire universe. This semantic and logical abuse is already not founded for a basic holographic encoding, since the definition or the grain of it is reduced at each of its extractions, although it retains its structure thanks to the properties of Fourier's transformations. The mistaken idea that any point of the holographic encoding contains the entire hologram, which is nonsense which does flores...

However, the holographic interference concept has similarities to the IBOSDSOO cosmological substrate model of Oommo, which is non-local and having an almost infinite number of dimensional axes, driven by standing wave interference. So, the standing wave interference would be the "hologram" where emerge masses, volumes and forces or energies manifested in the physical world...

The holographic fashion led to a multitude of interpretations, not necessarily consistent among them. For example, in 1994 the New Zealand physicist Gerard Hooft expresses this idea to designate a three-dimensional universe emerging from two underlying dimensions...

In the 2010s, physicist Nassim Haramein develops a "theory of the holofractographic universe" and incorporates many existing physical and philosophical concepts, including the "quantum potential field" of David Bohm and Einstein's equations of the "unified field theory". Nassim Haramein develops with innovative geometrical aspects of a double torus pattern and a fractal approach. It thus takes ideas around David Bohm, but restricting them around the concept of "quantum vacuum" and "white holes-black holes" where David Bohm saw a "dynamic holographic structure". Furthermore, the fractal view of the universe according Nassim Haramein seems an abstraction far to find phenomena that could be explained concretely...

Although Nassim Haramein has realized that conventional physics was obsolete, it tries to extend the work of Albert Einstein and fails to produce a model nor smarter nor more insightful than that of David Bohm. The thesis of a non-local subquantum substrate of David Bohm remains most relevant to date, despite its shortcomings by its lack of cosmological vision...

Thus, according to our OOMMO friends, the universal substrate would consist of "deca-dimensional axes interconnecting nodes" (an almost infinite number of axes in reality, 10 of which are sufficient to express the world as we know) named "ibosdsoo "which each axis is named according to OAWOO Oommo terminology. Instead of Strings that vibrate according to generate space, matter, energy and forces, these 'multidimensional nodes' would be 'virtual ball' (mathematically décadimensionnel) which can operate angular rotations under different axes. Unlike the Strings are assumed to have a physical existence, these 'ibosdsoo' would be only the result of interconnections at least 10 mathematics dimensions. The ibosdsoo does not exist in itself. The ibosdsoo exists only in relation to another ibosdsoo. The IBOSDSOO substrate is non-local and as Niels Bohr had foreseen, it does not exist as it is not shown.

Two ibosdsoo are associated with a very small angular difference. They are then to support the manifestation of all matter, energy, space, time, gravity, electromagnetic or nuclear forces.

They form a pair of ibosdsoo named 'ibosdsoo-UU'. The ibosdsoo-UU is the universal substrate of all matter, all energy, space or time in the Oommo cosmological model, and a priori the vast majority of exo-civilizations in the cosmos, with diverse formalizations.

The IBOSDSDOO-UU associates themselves and form chains following their different axes. They make up, in a way, the theoretical grid supporting any manifestation of force, time or space in the cosmos/anti-cosmos.

In a certain way these knots precede what we call dimensions and they integrate themselves in a theory of the twin cosmology, essential in order to contemplate interstellar travel. Depending on their angular rotation, they can manifest various aspects, various natures and modify the very state of matter.

PAIR OF IBODSOO-OU (IBOZOO UU)

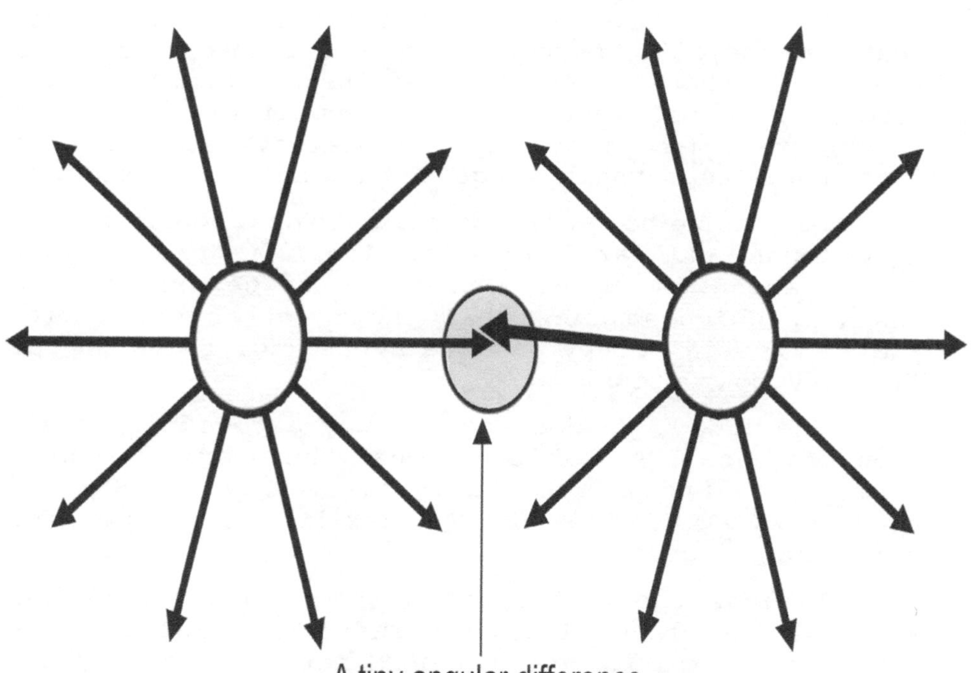

The example of the dragonflies:

The universe 'is like a swarm of dragonflies' whose wings are at various angles.

All these dragonflies flutter about in such a way that not one of them shows an orientation of its wings similar to that of any other of its sisters. In other words, there will not be a single pair of dragonflies which, at a given instant, will overlay each other in such a way that their wings and abdomens coincide.

But, as we have already said, this image is extremely rough and far-fetched in its analogy. Firstly, each dragonfly occupies a place in space during each instant. Which implies that their center of gravity and of inertia occupy definite areas (according to our illusory conception).

One IBOZOO UU does not occupy any definite position, we cannot say about it that there exists a probability to find it located in a given point. However the IBOSDSOO UU IEN AIOOYAA (exist). (IEN: meaning pair, two)

Besides that, this flying insect has a mass and a volume (at least for our mind). The IBOZOO UU is not a particle possessing a MASS nor is it corporeal. In a first conceptual approximation we could describe it as a bunch, or fascicle, of oriented axes. What is paramount for such a cluster are precisely the angles formed by these axes, rather than the axes themselves which are similar to a mathematical fiction.

To be more precise: what we might call INFINITESIMAL TIME INTERVAL is but the outcome of an angular orientation difference between two linked IBOZOOs or IBOZOO UU.

If after this rough explanation of our Theory of Space, you think that space is a dense mass of particles similar to atoms, you are mistaken. The reason is that, the particles of a gas such, as you know, it takes up probabilistic positions within an enclosure, whereas this is not the case with the IBOZOO UU.

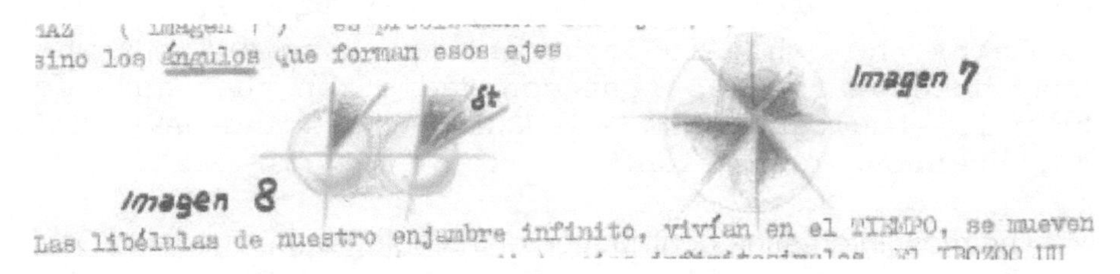

Also you cannot equate such a space to the antiquated concept of ETHER banished by the theory of relativity, since the IBOZOO UU network is not in any way an elastic environment in which the atoms of bodies might be immersed. You could also ask us: in relationship to which AXIS of universal reference are the angles of the IBOZOO UU orientated?

The answer is, naturally, WITH NONE. There isn't one single axis of reference in the WAAM (bi-cosmos) since that would imply imagining a real straight line in the Cosmos. Yet, such a straight line, as we have indicated is a mere fiction.

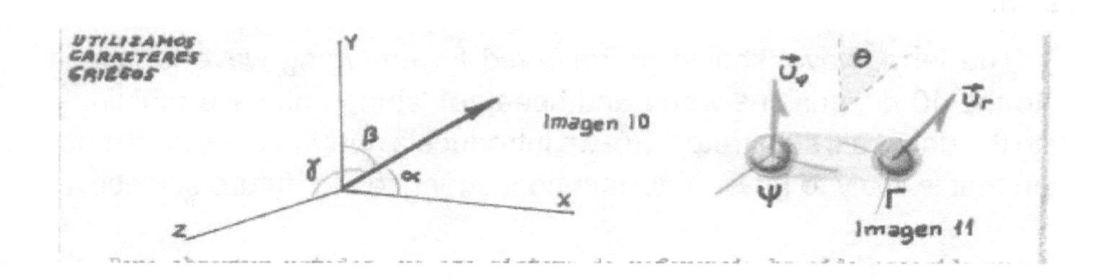

When now we refer to the angle taken by one of the imaginary axes of an IBOZOO UU, we are referring to any other IBOZOO UU chosen conventionally as a model of reference. [excerpts from document D59]

WAVES AND IBOSDSOO

D45: 'But do not forget that we consider the cosmos as a decadimensionnal system, WOA generates an infinite series of wave trains (sinusoids functions) of different frequencies, amplitude and phase. SPACE is crooked, causing a series STATIONARY WAVE and NODE which are reflected in the infinity of WAAM. These standing waves are only the folds of SPACE-TIME CONTINUUM we call MASSES (Galaxies, Gas, Pet, etc.). This explains the confusion of terrestrial scientists when they observe the apparent contradiction that an electron is simultaneously CORPUSCLE (mass) and WAVE. It's an ingenuous confusion.'

The most interesting contribution of the holographic model of David Bohm, is to present an universal wave substrate and non-local from which emerge the particles, very close in fine from IBOSDSOO model.

It is these wave trains crossing the universal substrate, form chains of IBOSDSOO.

These waves, channels or strings of IBOSDSOO, bringing forth time and volumes that constitute the space-time, as in particular the strengths and the masses as particles...

Regarding the 'wave trains' of the universal substrate, I think we should introduce an important semantic distinction. Indeed, the universal substrate being non-local, speak of 'waves' probably does not make sense ... that's probably why our Oommo friends rather speak of 'strings' ...

The term 'wave' should be reserved for emerging wave functions into the 10 dimensions world and speak of 'strings of wave functions' into the universal substrate... So, we introduce a new conceptual distinction that is likely to have important consequences for future science...

THE UNIVERSAL DIMENSIONS OAWOO

Cosmos/anticosmos se définirait mathématiquement par un minimuThe cosmos/anti-cosmos can be defined mathematically by a minimum of 10 angular 'dimensions' or OAWOO according to Ummo terminology. These minimum theoretical '10Ds', are present in each and every cosmos pair. The meaning and nature of these 10 angular 'dimensions' is not easy to grasp. They are quite foreign to our ways of conceiving time and space (D41-15):

> 'Our cosmos is what you call a space-time continuum (it requires 10 dimensions to define it mathematically). We could speculate by assigning an infinity of dimensions to it, but this is not something we are in a position to prove.
>
> Out of these ten dimensions, three are perceptible by our sensory organs and a fourth—time—is perceived psychologically as a continuous flow going one way along what we call UIWIUTAA (arrow or orientated direction of time).
>
> You can imagine that our primitive bi-cosmos was rather like a small empty sphere. A small universe without galaxies, without intergalactic gases, only space existing in time.
>
> Each new curvature implies one dimension and finally, "pleats" or "folds". Observe that we are using comparison, a symbol, because this can be expressed adequately only in mathematical terms. For example, the expression "pleating space" might seem childlike, but it is very didactic.
>
> ... Having arrived at this instant, the entire universe is reduced to a network of IBOZOO UU, all the components of which are orientated at a null angle (zero radii) which, if we could perceive it, might seem like a point with an infinite mass density (this has been well understood by your cosmologist brothers of the Earth and is absolutely certain).
>
> What is not certain though is that this "cosmion" or primordial universe, is unstable and might consequently explode. If the adjacent Universes did not exist and if there were no more than two types of masses (and not four) which would be disrupting this hyper mass by unbalancing it, this would be the final stage of the cosmos described here. There occurs then an accelerated expansion through the initial energy input of this disturbance (which is inversely proportional to the radius).'
>
> "

Semantic analysis of OAWOO

D59: "any particle, (ELECTRON, MESON, or GRAVITON) is PRECISELY an UU IBOZOO oriented in a particular way in relation to others.

For us, RIGHT SPACE does not exist, as we explain below, and the CONCEPT OAWOO (DIMENSION) takes a different meaning for us. Such dimensions are not related to SCALAR QUANTITIES but ANGULAR SIZES.

We consider in the sphere of figure S59-f10 an OAWOO (with this name we shall specify in the sphere as well the concept of AXIS of terrestrial mathematicians, as the VECTOR with its attributes of module, origin, and extremity). In this case, you will translate OAWOO by RADIUS VECTOR U (U arrow)"

"The OAWOO, on the other hand, IS NOT A CONVENTION: This is not a simple parameter, an arbitrary way of representing and IBOZOO UU (such as perhaps, for example, the number of leptons invented by the physicists of the Earth).

The OAWOO does not exist without imagining it connected or related to another OAWOO with which it forms an ELEMENTARY ANGLE which we call IOAWOO.

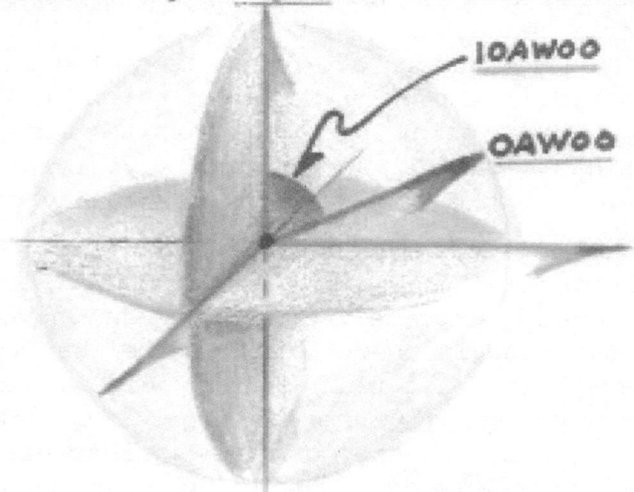

IT IS NOT POSSIBLE TO CHOOSE a reference system IN THE SAME IBOZOO UU. Such a REFERENCE SYSTEM MUST BE MADE BY ANOTHER arbitrarily chosen IBOZOO UU. And it is precisely this IOAWOO Theta (ANGLE-DIMENSION) which gives the IBOZOO UU all its transcendent meaning.

In fact, it will be much more accessible to imagine the concept of IOAWOO (we translate this as 'ANGLE FORMED BY TWO OAWOO'). You will recall the previous documents and how we identified this IOAWOO with some familiar measurements (LENGTH AND TIME)."

D59: "We, on the contrary, know that the WAAM (cosmos) is composed of a network of IBOZOO UU. We conceive of SPACE as an associated set of angular factors...

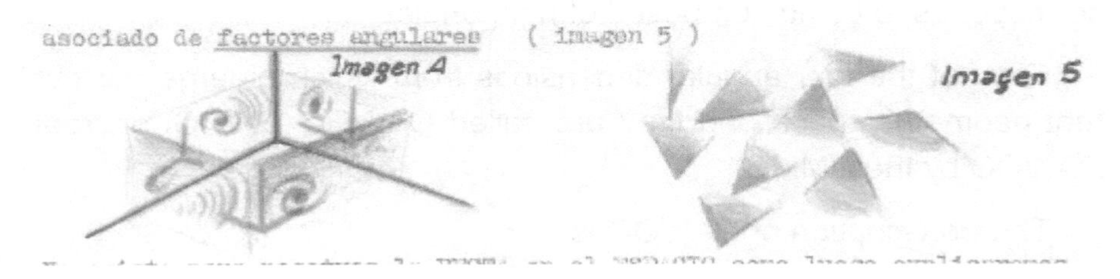

RIGHT line does not exist in SPACE

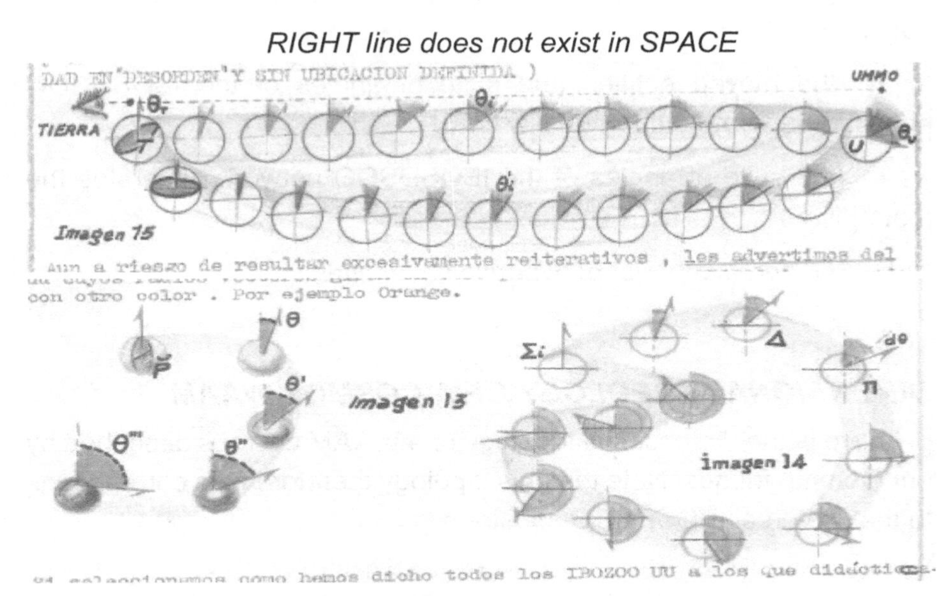

The concept of OAWOO indicates the axial orientation of an IBOSDSOO, where each IBOSDSOO is constituted by a 10 angular dimensional OAWOO.

An angular dimension is constituted by a "bunch" of a "physical infinity" of OAWOO. Each of them is separated by an elementary angle IOAWOO, each angular dimension covering 360 degrees (0 to 2pi).

Two "successive" OAWOO of a given angular dimension thus forms and elementary angle (ultimate, minimal and incompressible) IOAWOO.

As the semantics indicates, every IOAWOO identifies every pair of OAWOO in a unique way, from which emerge the material entities of space, time, mass, and energy.

The angular dimension (or bunch, or angular factor) is bounded and limited by two OAWOO which is orthogonal in the mathematical sense of the word (within an angular dimension, the IOAWOO covers 360 degrees). In the particular case of tridimensional space, orthogonality corresponds to being composed of right angles.

Each of the four angular dimensions from which emerge our current geometric space and time are called OAWOO UXGIGI or "real" OAWOO by the UMMO.

The transcription of OAWOO is:

- [(O) entity [(A) movement [(W) generation "has" (OO) material entity]
- the moved entities (the suite of angles of the IBOSDSOO network) generate the materiality
- the suite of angles of the IBOSDSOO network generates the materiality
- *Generative "Axis" of dimensional entities*

DIMENSIONAL TOPOLOGY OF A COSMOS WAAM

Here is the dimensional topology of a WAAM cosmos described by our Oommo friends. Note that this topology therefore also corresponds to the universal IBOSDSOO substrate.

Our mathematical modeling waam-waam tetrahedral, only requires 12 dimensions for is expression.

our functional physical model only considers 10 dimensions:

The dimensional coordinate system representing time (t) is reduced to a single axial dimension around which pivot the 3 others trihedrons.

In each of the other 3 trihedrons, each dimension is defined relative to the angular axis T.

the angular positions of the dimensions are separated by a minimum angular increment, verified experimentally about 6.10 ^ $^{-11}$ Radians.

Below this increment, dimensional vibrations merge into a single harmonic.

There is therefore in practice only about 10 ^ ^{11} different angular orientations between the axis and a dimension t in the range from 0 to 2pi in each degree of freedom.

Each combination possible directions through 9 free dimensions are a waam (universe/cosmos).

The number of possible waam is thus limited to a maximum of 10 ^ ^{495}.

The waam-waam is limited. Also, it is limited the potential of emergence within each waam.

Each waam, including our cosmos, except 2 cosmos limits, is expressed as 10 dimensions, not all seen by the oemii (human body).

Each dimensional trihedron contains 3 dimensions.

You can represent each trihedron in the form of a pyramid with a triangular base, whose edges are elastic and articulated at each vertex according to 9 degrees of freedom, one of the vertices being further articulated about the axis T.

From each of 3 free trihedrons, edges cannot take the same direction as any other including, in particular that of the axe T.

In other words, 6 degrees of freedom: height, width, depth and three-roll axes, pitch and yaw angle.

The other additional 3 degrees of freedom must be related to time.

Here the geometric representation that we make (achieving Philippe Douillet), by convention:

- *the blue trihedron is the mass dimensions (with a white axis mark)*
- *the red trihedron the volume dimensions (with a white axis mark)*
- *the yellow trihedron forces dimensions (with a white axis mark).*

The time axis is the central axis in black.

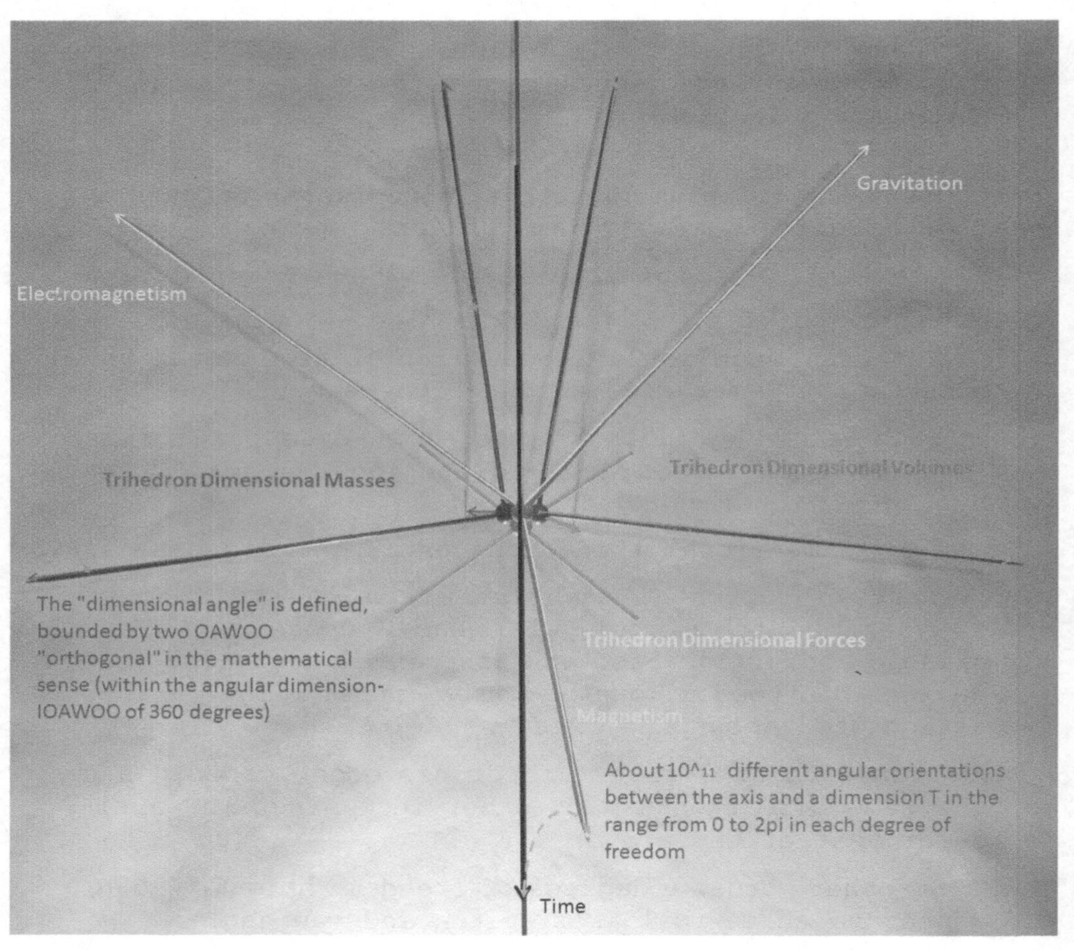

Remember that each dimension itself contains an almost infinite number of angular orientations (approximately 10 ^ 11 angular orientations) ... For example, if we look at a material object in our space-time so we see is apparent already 9 general dimensions.

In the volumetric dimensions can manifest speed and acceleration, linear and angular. What makes us 12 angular orientations.

In addition, mass axis of the object, there are axes of the energy which contains (potential, kinetic...) and the forces that can be applied to it on three sides according to the three axes...

If the object can be deformed, we will have angular orientations as the expansion or compression in the 3-axis or the bending and torsion in the three angles.

The object can respond to electromagnetic waves, heat, light and/or radiation, must be added the axes related to transmission coefficients, reflection and absorption of these different range of wave...

If the object is soft or liquid, we still add axes to the fluidity and viscosity. Thus, a material object is described by an almost infinite number of angular orientations...

The OAWOO and the GAUGE GROUPS

The analogies between our usual vector dimensions and angular dimensions are not easy. In "Presence, UFOs, Crop Circles and Exocivilizations" we made analogies between angular dimensions (or OAWOO) with groups and gauge symmetry. Ultimately, each dimensional trihedron, Force, Volume, Mass, would be a group and gauge symmetry. Whatever the angular orientation in the trihedron, Force, Volume or Mass remains invariant.

> "Any particle (electron, boson or graviton) is precisely an IBODSOO UU orientated in particular way with respect to others. (D59)"

> "Remember that the vectors representing the gravitational, electrostatic and magnetic fields form a trihedral in multidimensional space. The three fields are in fact identical. It is our illusory physiological perception that attributes to them different natures depending on their orientation (D57-3)"

THE MULTI-BIG-BANG OF THE WAAM-WAAM

The multi-cosmos concretize the physical realization of the potentialities of metaphysical or transcendent "objects".

"So the WAAM-WAAM is organized as and when it is generated by WOA. This process is both simultaneous and infinite."

The researcher VMo gives us an educational vision. This, however, does not reproduce the complexity of the organization that one can imagine in 10 dimensions.

The coexistence plan can be interpreted as the existing set of IBOSDSOO UU of all the universe substrate.

Zones in vertical U suggest the expansion of the previous universe pairs.

Thick lines denote a linear dimension, past each consisting of the cosmos, after which this continues to condense the past. These arrows show how VMo includes opposite time.

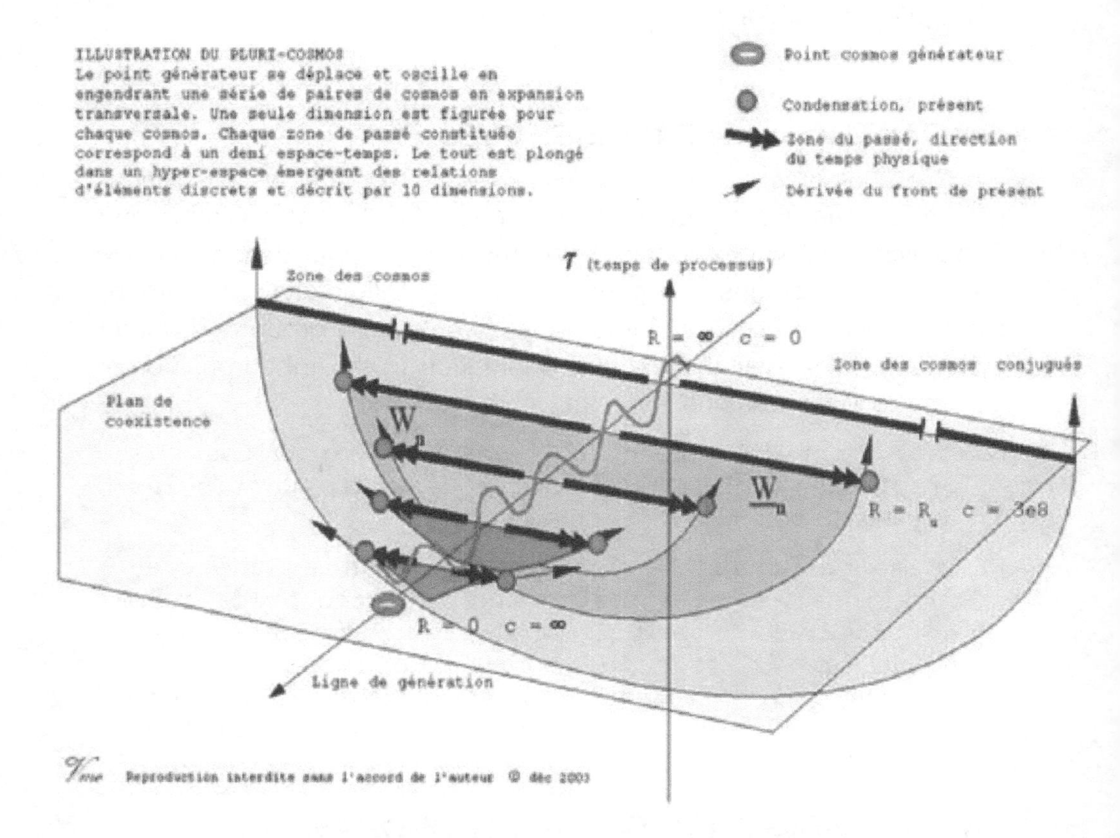

It is suggested that the extreme universe with zero radius is the initial point of the periodic generation of a new pair of cosmos.

The vertical axis is a time of change outside of physical time without which we can hardly understand the existence of the change, regardless of our usual concept of time (a kind of time for construction activity in eternity. From the point theoretical view for VMo it's a stochastic process time).

The Big Bang of our cosmos and the **OAWOO**

For each bi-cosmos, cycles made up of a Big Crunch followed by a Big Bang occurs. At the point of balance between the Big Crunch and the Big Bang, each bi-cosmos is reduced to a chain of "multidimensional knots" among which the 10 axial dimensions are "aligned" without the slightest angular difference which would make the manifestation of "something" in its midst possible. Each axial dimension is uniform to the infinite. Time is reduced to one single infinitesimal unit, in other words, it does not yet really exist. Similarly, the dimensions of space are reduced to something like a kind of dot and the dimensions of mass are therefore all concentrated in the same way in quasi-infinity. The quasi-infinity of IBOZOO which constitutes the substratum of the bi-cosmos is equal to itself, and it manifests by a kind of dot.

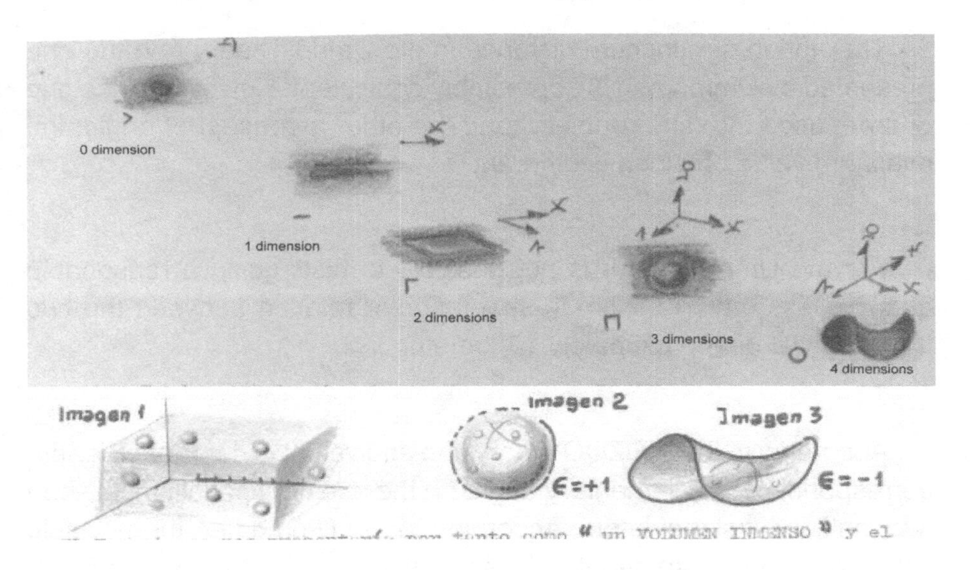

Three spatial dimensions, four space-time dimensions into a negative hypersphere

The birth of Time

The first axial component which manifests gives the orientation of the other angular "dimensions" or OAWOO. Thus, in each bi-cosmos, time is born. The unit of time has the same value in each pair of bi-cosmoses. The minimal amount of time corresponds to the smallest angular variation or IOAWOO of this axis, which exists as a discreet value, so that time is discontinuous and finite. What we define by the time of Planck 5,391 x 10^{-44} seconds) might correspond to an elementary angle on the axis of time, between two "multidimensional knots" IBOZOO. It corresponds to a "1D" angular dimension.

The birth of Space or of "Spatiality"

The anisotropy, the birth of space or "spatiality" starts from the Planck wall at 5,391 x 10^{-44} seconds. We can think it is also in this phase that the anticosmos appears.

Concerning the spatial dimensions, we can also draw an analogy with our usual vectorial approach. Planck's length of corresponds to the minimum diameter of a string in the String Theory: lp = 161,624 x 10^{-35} meters.

The notion of minimum distance in the String Theory and the one presented by Ummomen is appreciably equivalent. However, since one of them uses the "string-object" and the other a concept of angle, the minimum value obtained is different.

For the Ummomen, it is not possible to distinguish a reasonable quantity less than 12^{-13} cm in size (angular relation between the two IBOZOO UU of approximately 10^{-16} meters).

According to the IBOZOO theory, the equivalent of the Planck length corresponds to an elementary angle on the axes of spatiality, between two multidimensional knots. According to the Ummomen the possible value of the minimum angular distance is around 10^{-16} meters or 6.10^-11 radians. These angular spatial dimensions are tridimensional or "3D".

"The angular positions of the dimensions are separated by a minimum angular increment, verified about 6.10 ^ -11 radians experimentally."

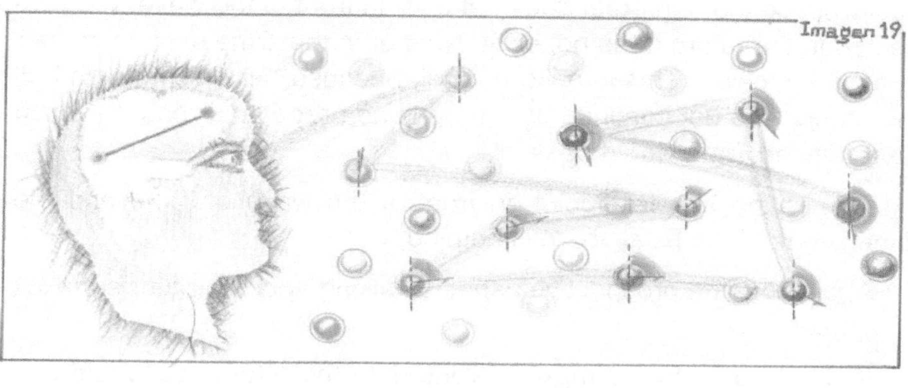

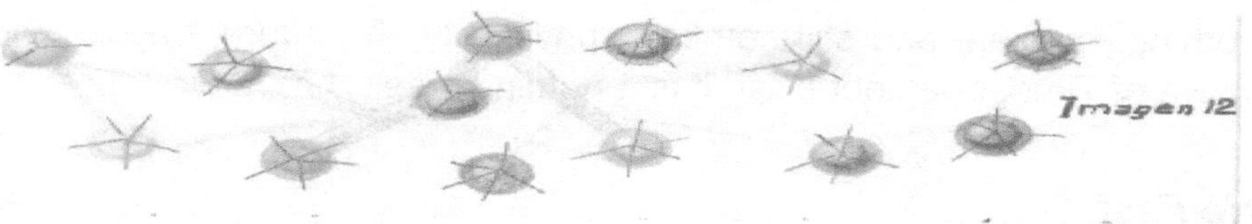

We can calculate that at an angle of 6 x 10^{-11} radius, to find a length of 10^{-16} meters, we must be at a distance of about 1.66 x 10^{-5} meters from the center of the angle (a very small distance that lies in the infrared).

The Thickness of Time

The Wave of Time

Some readers will find it strange that I speak of the thickness of Time. Let's go back a little bit to what leads us to this notion. Time orients all the other dimensions of our cosmos.

Time is like a speedboat on the sea of IBODSOO, that sea which is a non-local substrate from which emerge the dimensions of Time, Volume, Mass and Forces.

The speedboat of Time produces a wave, a wake, which is none other than the emergence of the dimensions of Volume, Mass and Forces.

No Future

As in this analogical example, we can clearly see that the wave of Time «carries» the other dimensions that constitute our Real. We easily notice that this Real does not exist in front of the wave. We suspected it, the future does not exist. Neither in the Time dimension, nor in any dimension. This is quite normal, because our Cosmos, and all the others, are not continuums. There is no predefined road on which a predefined Time would pass...

This is the simplest point, in front of the wave of Time and the dimensions of the Real, there is nothing.

Apart from the prospective aspects, talking about the future makes no sense.

If, like in a fantasy film, we «went» into the future, we would find nothing, nothing, and still nothing... physically, the future, beyond the Wave of Time, does not exist, it is a nothingness...

No Past

The same is true of the Past. Here again, the myth of the Space-Time continuum has wreaked havoc on our imaginations. Our senses and intellect are deceived by this false idea of the continuity of Time. We see a historical remnant of the past and we think that Time is a continuum. Whereas the remnants we see are nothing but the present time. This time is discrete in nature, that is, it is made up of small, distinct units, like the pixels in a photo, giving us the impression that the image is uniform...

However, it is true that the data of the past is contained in the Metabrain BB, but it is not physically possible to go back to the past, like in a fantasy movie for the good reason, that after the wave of Time, there is nothing either!

Again, to speak physically of the Past, after the emergence of the dimensions of the Wave of Time, makes no sense, because to «go» into the Past would just be to fall into nothingness...

Thus, before and after the wave of time that «carries» the other dimensions that constitute our Real, there is nothing, nothing but the nothingness of a non-local substratum. The wave of Time carries away the dimensions of the Real.

This is why the question of «The Thickness of Time» is important.

The calculation of «The Thickness of Time»

As we will discuss later, travel in other cosmos is done by tilting all the axes of the machine. This prob-ably requires having a common dimensional plane with the target cosmos, having various frames of reference between the 2 cosmoses. Especially when the cosmos do not have the same Time. In this case, the machine will have to keep its Earth Reference Time. We know that this must be calculated at 10^{-9} seconds to calculate a return to Earth of the machine in our Time wave, to return to the middle of the Time wave of our cosmos.

In fact, an accuracy of less than 10^{-9} seconds would send the ship into Nothing. He would remain stuck in his time bubble, outside of the Wave of Time. So that's what happens for a precision of 10 -8 seconds, we are out of the Wave of Time.

So there is a gap of 10×10^{-9} seconds to stay in the middle of the Time Wave. The gap will be the same on each side, so a total gap of 20×10^{-9} seconds.

Knowing that the speed of light is $0.3 \times 10^{+9}$, the thickness of the Time Wave is 20 x 0.3 = 6 meters.

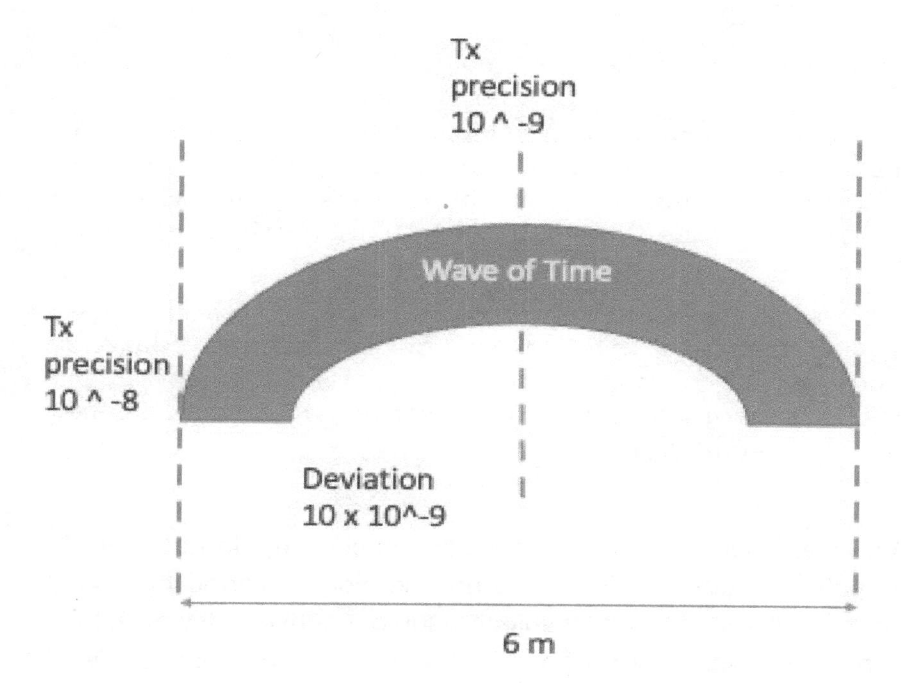

The masses and imaginary masses

Contrary to what we commonly think, the masses are not strictly related to volumes. But it is only when these masses emerge in volumes that forces can then be exerted on them. This is the case of gravity, which is exerted on the masses in volume creating a weight. Classical M + mass manifests as a "hollow through a fourth dimension Vector" and the mass—M would be manifested by a "bump" in the same vector dimension.

However, there are masses without volume, unknown to terrestrial physics of the XXI century, called by our OOMMO friends, imaginary masses.

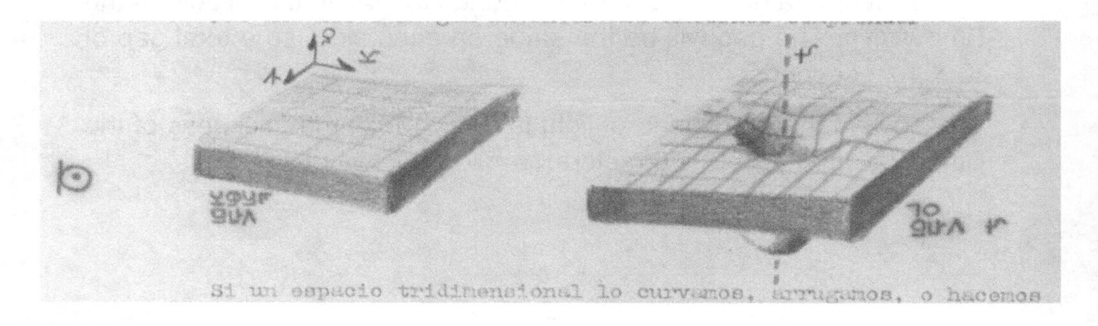

"If the adjacent universes did not exist and if there were but two types of masses (and not four) which were disrupting this hypermass by unbalancing it, this would be the final stage of such a cosmos." (D41-15)

"If we curve tridimensional space, if we bend it or if we create a kind of hollow in it (see figure 2) through a fourth dimension, this curvature represents what our sensory organs interpret as a mass (a stone, a planet, a galaxy)."

As far as imaginary masses are concerned, they do not seem to have spatial dimensions. They are not therefore distorted by spatial dimensions and are not perceptible through our senses. However, they relay the gravitational forces produced by the masses +M and—M of the anti-cosmos.

The two types of imaginary masses + √-M and—√-M make up the XOODII (cosmos and anti-cosmos relay layer). The gravitational forces produced by the masses +M and—M of the anti-cosmos, are transmitted to the imaginary masses and the imaginary masses transmit gravitational forces in our cosmos.

"... The singularities of one of them (concentrated ±√-M masses influence the adjacent universes (without ± √-M masses) ... The disturbances between cosmoses are produced because in one of them there is a type of mass which you might qualify mathematically as IMAGINARY (in another framework of the tridimensional beam).

This imaginary mass's speed "at rest" (maximum energy) is the speed of a cluster of electromagnetic energy (photon) ± √-M. The existence of this mass enables the interaction or mutual action, between Universes, although the imaginary mass involved is only located in one of the members of the couple." (D731)

The "dark matter"

In some areas of our cosmos significant and mysterious gravitational effects were measured. These gravitational effects are equivalent to a mass which represents 90% of the known mass of the cosmos. But it's impossible to detect the existence of this huge amount of invisible matter. Thus was born the hypothesis of a "dark" invisible matter which only the gravitational effects are detected.

From the Oomomen's cosmological model, we can think that "dark matter" would be the gravitational mass effect, of matter—M, from anti-cosmos onto the cosmos.

This 3D representation of "dark matter" according to the measurement of its gravitational effects, would be in fine, the representation of the +M and—M of the anti-cosmos UWAAM, the gravitational effect of which is relayed by the "relay" layer XOODII.

3D representation by Richard Massey

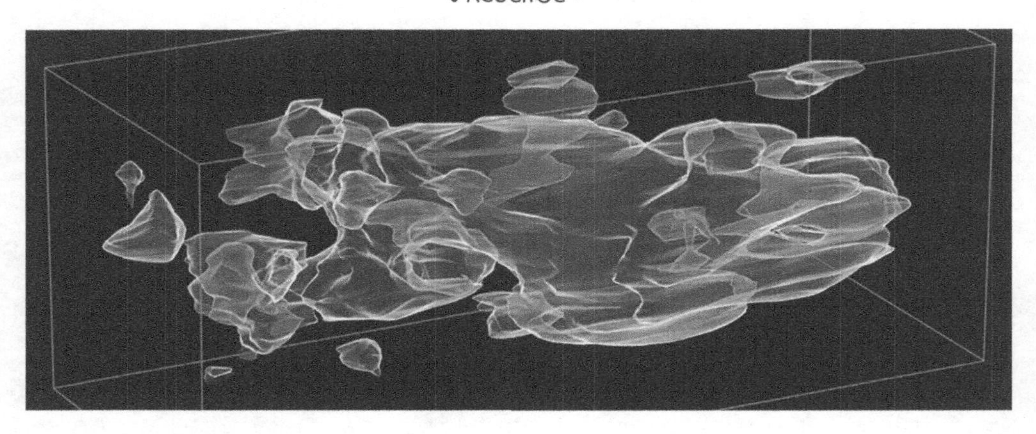

The "dark" energy

We think that the movements of huge masses of anti-matter inside the anti-cosmos have a gravitational effect on our cosmos through the cosmos-layer XOODII. It would have an overall effect of speeding up or slowing down the rate of expansion of our Cosmos.

The "dark" energy must be understood as the energy transmitted via the cosmos-layer XOODII. It is probably this effect which has also been interpreted as the energy of the quantum vacuum.

The switching of angular axes

The general principle of switching angular axis is to change the orientation of all IBOSDSOO of a given axis, called OAWOO. In other words, it is a permutation of gauge groups. The use of this principle in interstellar flights is described in *"Presence, UFOs, Crop Circles and Exocivilizations".*

A priori, there are an infinite number of axes OAWOO and therefore an infinite number of possible types of failover... We can assume multiple cases:

- The switching axes staying in our three-dimensional framework
- The switching axes staying in our cosmos
- The switching axes by changing cosmos
- The switching axes into the anti-cosmos

D68 "Matter under pressure somewhat greater loses its atomic structure is already known to you.

A pressure of 16 million atmospheres, (15,445,660 atmospheres AADAGIOOU call us [can be translated as critical pressure] simultaneously reverses all atomic particles [IBOZOO UU] The mass becomes PART OF ENERGY. The expansion that follows is immeasurable [This was the initial pressure of the whole mass of the Universe]."

NR22 "When we travel in our spacecraft in another dimensional framework, the telepathic link with OUMMO is possible if we do not change the time reference, the angular change in particle—OAWOOLEIIDAA—then being axially to the time dimension by processing similar to the orientation of the three spatial components and of the mass associated with three components."

The switching axes staying in our three-dimensional framework

The switching axes staying in our three-dimensional framework can have different effects on different axes that are failed over. We will see several examples of these effects onto volumes and masses in the chapter of communication with "spirits" which deals with "paranormal" phenomena.

In our dimensional framework, it is often the case of partial tilting the OAWOO axes. In particular, the axes of the masses—M, √—M and √ +M which are partially switched. The tilt angle is probably less than 90 ° in this case. The axis of the mass +M and spatial volume axes remaining invariant, positioned in our dimensional framework.

This type of partial tilting OAWOO axes described in "*Presence—UFOs, Crop Circles and Exocivilizations*" with the example of the SEG (Searl-Effect-Generator) a device that can levitate, but without changing its three-dimensional framework. In Searl Effect the case, a difference of electrostatic potential greater than 1.4 million volts per linear centimeters, causes a LEIYO effect, a partial loss of mass, remaining in the current three-dimensional framework. We can imagine that the OAWOO masses axes undergo partial torsion remains below 90 °. The total failover of the OAWOO masses axes would result in a transparency of the machine fully "unmassified".

The switching axes staying in our cosmos

Such tilting is also described in “Presence, UFOs, Crop Circles and Exocivilizations” we named “spatial susception”. It explains how a UFO can achieve a pseudo-change of direction at right angles at high speed.

This axis change produces a three-dimensional framework change with tilting the axes of the masses and volumes—M, √—M and √ +M. At no point mass +M is converted to negative mass—M. The axes of masses +M and time stay invariant.

In this case, the machine does not change of cosmos, but only its three-dimensional framework.

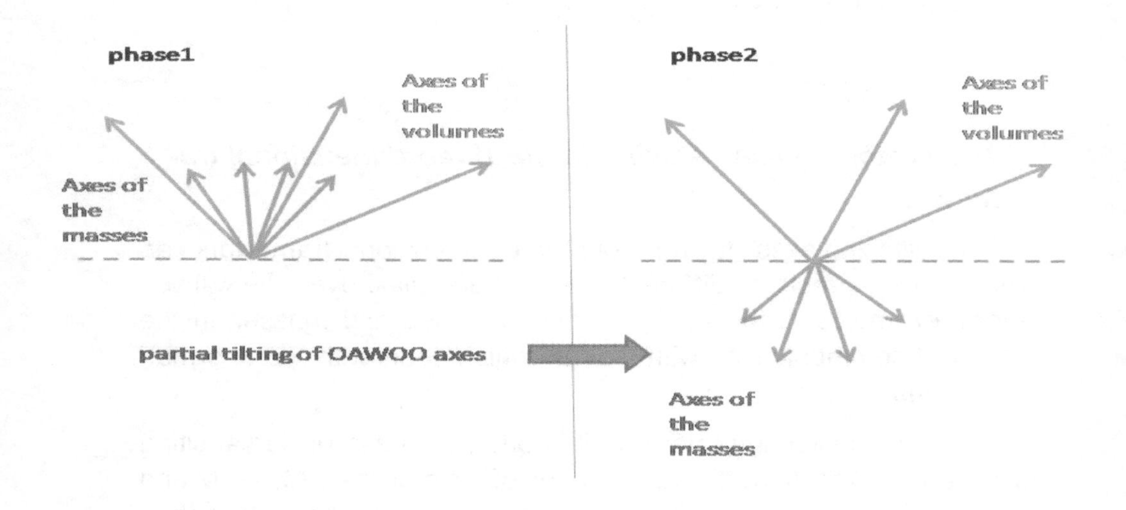

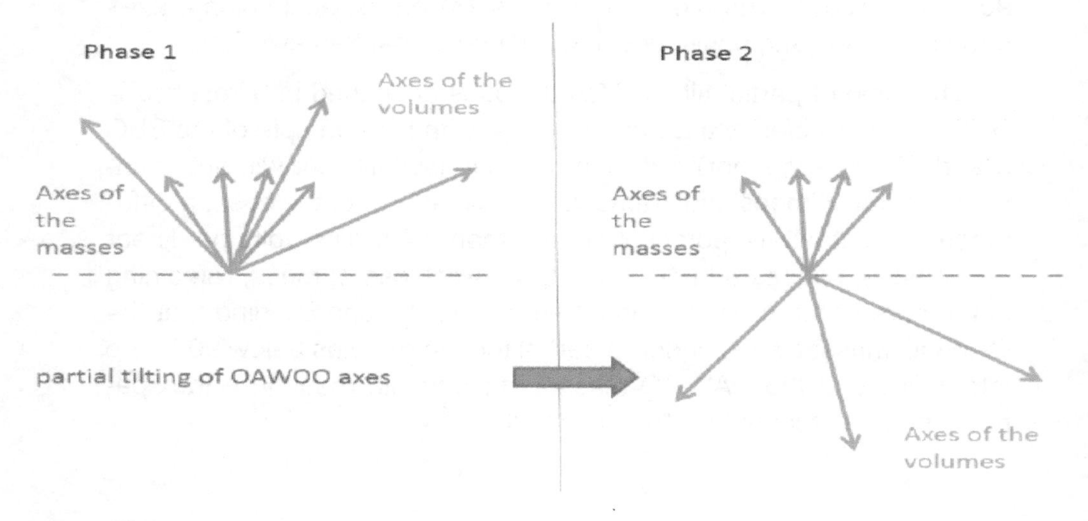

The UFO are equipped with a so-called device of "particles inversion", or more precisely "of switching OAWOO dimensional axes", called OAWOLEIIDA which directly converts the matter from the three-dimensional frame into energy into the other three-dimensional framework, and vice versa energy into mass. The OAWOLEIIDA concept concerns a "transformation of an IBOSDSOO UU network limited to the inversion of the three-dimensional axes of the IBOSDSOO UU".

The word OAWOLE can be translated by "the displacement [depending on the dimension-angle] generates the passage of [dimensional] entities from a physical medium to another." In other words, a rotation of 90 ° of the axial orientation of the "angular dimension" of the dimensional framework. The word IIDA can be translated by "Defines angular displacement".

To summarize, OAWOLEIIDA is "the displacement depending on the dimension-angle that generates the passage of dimensional entities by the substitution of an angular position".

In synthetic terms "the switching axis is defined by an angle"

(cf. *Presence 2, The language and the mystery of the UMMO planet disclosed*).

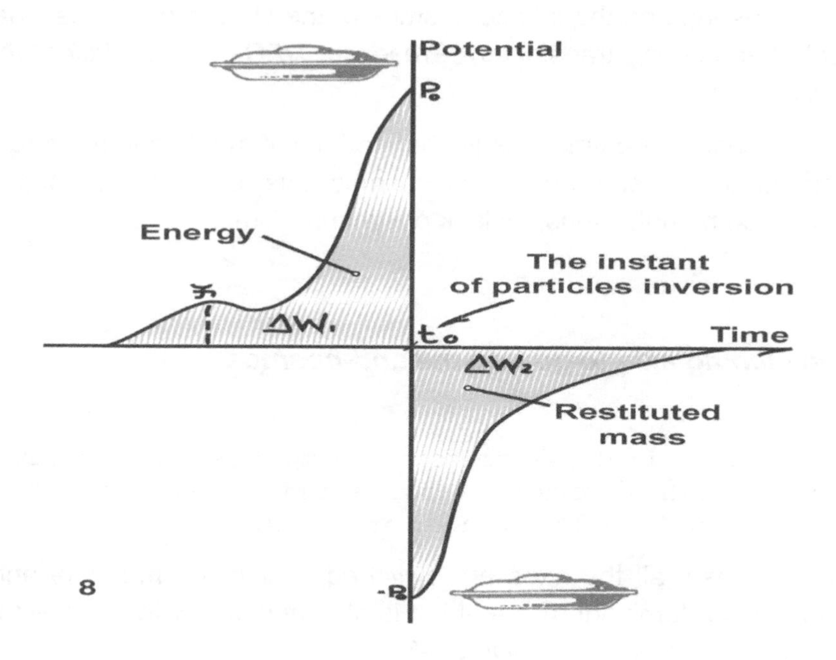

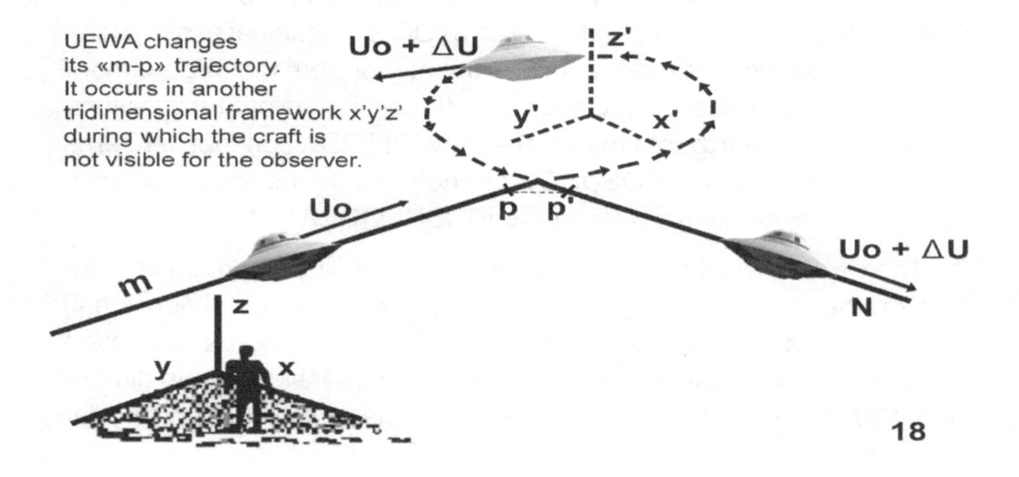

The switching axes with cosmos change

> D 731 "Although in some universe [cosmos]... In all WAAM [cosmos]..."

We can imagine that by modifying the X axis-dimensional OAWOO of an object; we can pass into another dimensional framework of another cosmos, among the infinite cosmos of the Universe. This is what we call "hyperspace transfer" in "*Presence, UFOs, Crop Circles and Exocivilizations*".

In this case, if the time axis is changed, i.e. if the time in the target cosmos is not the same as in our cosmos, this results in the loss of telepathic communications, by lack of synchronism...

The switching axis to go into the anti-cosmos

> NR22 "Sometimes, we are switching axes temporarily in UU-WAAM, reversing by rotating pi radians, all dimensional angles of subparticles."

In this case, all the axes are switched, including time. The anti-cosmos being dominant mass—M, it is also necessary to reverse the burden of current Masses +M into—M.

The case of black holes

The gravitational collapse of neutron stars accumulates a critical energy which combines the gravitational collapse of the mass.

Overall, the object collapses and "pierce" the XOODII to reappear in anticosmos with negative mass. The result of the switching is energy remains prisoner into the XOODII, as imaginary mass. In this case, the whole mass M becomes—M into the other cosmos. The energy becomes √—M, which is the imaginary mass in the other cosmos.

H2 UMMO:— In reality, the Black Hole is impossible, because when a star collapses, comes a time when it becomes neutron star. If it keeps collapsing, comes a time when it disappears from our universe with a "border effect".

The black hole ceases to exist and the universe reduces its mass of the collapsed star.

What you consider as black holes are actually neutron stars and not authentic black holes. For example, one of the first neutron stars is in the constellation Cygnus at 8,365 years-light. In reality it is a neutron star, not a black hole. The Schwarzschild radius appeared and disappeared instantly. It's instantaneous. At this moment the mass of the black hole is converted into NEGATIVE mass into the other universe. The mass has disappeared, as negative mass, mass with a negative charge that is to say, different electric charge. And energy of the black hole is converted into [imaginary] mass.

H7 UMMO: What you call black holes, neutron stars are. A "border effect" occurs. Whole mass M becomes—M into the other universe. The energy is converted into the square root of—M, which is the imaginary mass into the other universe.

For example, following the Schwarzschild radius formula should the sun has a radius of 3 km to become a "black hole" and the Earth should have a radius of 9 mm to do the same. The "black hole" does not really exist either, this means that they are simply supermassive neutron stars that occupy the heart of galaxies.

We can imagine that neutron star supermassive can be used by interstellar craft with a change in dimensional framework to take advantage of some accelerators displacement effects, enjoying a "gravitational airflow" into the switched dimensional framework without risking absorption by the supermassive neutron star...

TETRAVALENT LOGIC

The Ummo people use a tetravalent logic related to ontology, to cosmology, to primary phonetic concepts and language itself. Understanding this logic allows us to deepen and clarify the ontology of cosmological entities that we have described.

Summary of the four values of the tetravalent logic. *(See details in PRESENCE 2—The language and the mystery of the UMMO planet disclosed.)*

THE FIRST VALUE OF TETRAVALENCE

The first value defined by the tetravalent word AÏOOYAA (verifiable existence) is linked to ontological positions of the primary concepts "O" and "OO", and also to cosmo-physical concepts that we will now explain in detail.

The word AIOOYAA

The primary concept "A" is relative to the infinitesimal displacement of "angles-IOAWOO" of each IBOZOO because it is the basis of the Ummo physics. The primary concepts "AI" can be translated by the concept of "action".

There is a positive truth value for AIOOYAA when the network of a multidimensional IBOZOO shows 4 of these angular dimensions. Anything sized angularly in a space-time is characterized by the infinitesimal displacement of "angles-IOAWOO" of each IBOZOO chained in each of the angular dimensions. If there is no "IOAWOO" in a space-sized angle, there is no verifiable existence. The limit of what is verifiable in sized space is ultimate "angle" IOAWOO which identifies the link between two IBOZOO of a string, following an "axis" OAWOO:

- The movement identifies a materiality that has a dynamic spatiality

- The angular displacement identifies a spatial materiality

The infinitesimal angle "IOAWOO" identifies the materiality of things (4-dimensional angular spatial, temporal) from 10D. In other words, this is: "the action to identify things materialized in the space-time" of 4D from 10D.

The field of Cosmology/Physics

The documents express the concept of "truth" in our 4D cosmos (including time and space) "what is dimensional with characteristic of time and space."

```
D105: "the network of I.U. that is the AIOYAA [...]
from various perspectives."
```

Briefly, AIOOYAA is what may occur in the space-time proposed by Einstein/Minkowski. In other words, AIOOYAA is the network of IBOZOO 4D-angles (a UXGIGIIAM WAAM), which occur from angular-10D.

There is a positive truth-value AIOOYAA when an IBOZOO multidimensional network manifests 4 of the angular dimensions that is translated as "The action to identify the things materialized in space-time."

The second value of tetravalence

The second value of tetravalence is defined by the word AIOOYEEDOO, which means "no reality outside any verification".

The word AÏOOYEEDOO

The phoneme "edo" expresses the concept of "MISSING—NOTHING—FALSE." For example "YAEYUEYEDOO (amnesia of fixation)"

Since there is no negation in the Ummo language, here we have the concept "no memory". In another example, "ASNEIIBIAEDOO (absorption by the BB) or disappearances" we also clearly see the concept of "NOTHING—FALSE." We also see this in a third example, EDDOIBOOI (WITHOUT WORK).

Lastly, we see Nothing-False in the following quote "On NOTHING we assign a verb that has no meaning for you; AIOYAYEDOO". The word AIOYAYEDOO then is concatenation (series of linked ideas) which springs from the concept AIOOYAA, which is dimensional and means "NOTHING—FALSE".

The overall expression EEDOO translates to "something" missing:

- The model (or conceptual) has a material form
- Conceptualizing a material form

The translation of AIOOYEEDOO = Moving identifies a materiality that has a spatiality that conceptualizes the material form.

We have: "Action to materialize things in a space that conceptualizes what is missing (and is therefore unverifiable)".

In other words, the "Action to conceptualize the absence of things in space-time".

There is a truth value which is negative for AIOOYEEDOO the "Action to conceptualize the absence of things in space-time".

The cosmo-physical and primary concepts

There is a positive truth value for AIOOYAA when a multidimensional IBOZOO network manifests four of these angular dimensions. Conversely, AÏOOYEEDOO is the lack of expression of these four angular dimensions.

THE THIRD VALUE OF THE TETRAVALENCE

The third value of tetravalence defined by the word AÏOOYAOU (real phenomenological potential or partially known). In some cases the result is "TRUE" and in other cases the result is "FALSE". For example, a quantum phenomenon as the position of an electron is purely statistical. Sometimes the electron is there, sometimes not. This is what Oommo people call AIOYAU and they associate this value conditional existence.

The word AIOOYAU

We understand from the word AIOOYAA, that the "action to realize the things in space-time" is "dependent," i.e. conditional or indefinite. The word "a"—"i"—"oyaoo" expresses the idea of "truth that is conditional or determinable."

"Reduced to a phenomenological potential reality or partially known (A ∩ B)"

"This state AÏOOYAOU is well summarized in the paradox created by your Schrödinger which led to the deduction that two potentially contradictory statements could be superimposed due to the quantum nature of the phenomena implemented in the experiment".

The field of Cosmology/Physics

Entities that have a quantum nature have a "reality" which is conditional or indeterminate based on the observation of the quantum event itself. This cosmological/physical event which happens in matter is due to an angular displacement of the "axis" OAWOO which follows the quantum event. These entities have a materiality which depends upon a statistical value, so their materiality is indeterminable.

This is where the "Action to identify the things materialized in space-time is conditional (or indefinite)."

THE FOURTH VALUE OF TETRAVALENCE

The fourth value is defined by the word AÏOOYA AMMIE (unverifiable out of a field of individual or collective consciousness). La dernière valeur est très importante pour nous au quotidien. The last value is very important for us everyday. Feelings, emotions, our interpretations of 4D physical world exist in our heads, outside the 4D physical world. This existence is "TRUE", but only for US, outside the physical world 4D; it's the "TRUE" of our interpretations in our 4D physical world, our feelings, our emotions... No one but me knows what I feel or what I interpret. According to documents from Oommomen, all information from the neurological brain is also stored simultaneously in a cosmological receptacle usually call "soul". And every soul that Oommomen call BUAWA can store and produce, emotions, thoughts, which are converted in the brain to adapt to our environment. This is what Oommomen call AIOYAA AMMIE and they associate the value of an existence outside the physical world 4D.

The word AMMIE

In the Ummo culture an abstract concept such as a feeling or soul does exist outside of our cosmos, but it does "NOT EXIST" in the point of view of our 4D cosmos. Otherwise expressed the concept is: the non-existence in our 4D world, but an existence in another entity cosmological.

"First we distinguish between two classes of beings existing in the UAANM (COSMOS) in contrast to two other major genres" OF THINGS (SERES) NOT EXIST.

```
These are: AIOYAA AMMEIEE UAA [Such as WOAA
(Generator), BUAUAA (Human Spirit), BUAWEE BEIAEII
(Collective Spirit) or BUAUAA BAAIOO (Spirit Of The
Alive being)]

IN ADDITION, AIOYAA AMEIEE OUEE (Such as the content
of information, the sensation of pleasure, or a
folk tradition).

"True outside the WAAM [our universe]"

"AÏOOYA AMMIÈ (¬B ∩ ¬A), unverifiable out of a field
of individual or collective consciousness."
```

In the word AÏOOYA AMMIE, the primary concept applies to AMMIE AIOOYAA. In other words, AMMIE applies the concept of the real 4D cosmos. In the word AMMIE, the primary concept of movement (A) applies in an inseparable way to identify a concept (IE), which is AIOOYAA itself in this case. AMMIE is translated to be:

- Moving inseparable from the identification of the concept
- Move [out of our cosmos] inseparable from the identification of the concept [the concept AIOOYAA itself in this case]

The cosmo-physical and primary concepts

(See also Chapter The primary phonetic concepts). The primary concepts "E—Concept" and "EE—model" are related to the fields of cosmology/physics. The Oomomen speak of two other major genres "OF THINGS (SERES) NO EXISTING", for which the primary tetravalent values and concepts are:

- The primary concept "E—concept" may be associated with any "concept" and in particular to the following concepts: AIOYAA AMMEIEE UAA designating cosmological entities that exist outside of our cosmos. Such as: WOAA (Generator), BUAUAA (Human Spirit), BUAWEE BEIAEII (Collective Spirit) or BUAUAA BAAIOO (Spirit Of The Alive being). These cosmological entities are not assignable to the primary concept "O" of 10D multidimensional entities of our cosmos that include features of time and space. Those entities have no time and for us they can best be seen as concepts.

- The primary concept "EE" (meaning model) may be associated with the tetravalent value AIOYAA AMEIEE OUEE (Such as the content of information, the sensation of pleasure, or a folk tradition). These entities are linked to patterns in the cosmos called BUAWEE BEIAEII (meaning Collective Spirit), and designated by the term "BB".

That is: "Action to identify the things outside of our 4D space" in other words "something real in another cosmos".

Excerpts of Cosmobiophysics

The Ummo people conceive the universe as a being aware of everything. This form of Theocosmology is based on a science called "Cosmobiophysics". So their perspective, their "spiritual" or "religious thought" is scientifically based in all its aspects. We selected excerpts from their texts that best express these ideas...

"It is certain that an early period of history, UMMOWOA (a prophet like Christ) appeared among the inhabitants of UMMO surrounded by a mystical aura. WOOAYII UMMOWOA could mean something like "divine UMMOWOA" although he did not develop an institutionalized religion around its memory as it happened on OYAGAA (planet Earth).

Modern Cosmobiophysics throws enough light on this magnificent phenomenon which is based on sharp cosmological laws. Emotional connotations and biophysical interpretation of this event is far from the notion that you have prepared on the historical fact of the birth of Jesus (which, as we will explain, is similar to UMMOWOA).

For you, the figure of Jesus is "divine" and surrounded by mystical-religious connotations. It would be a fact supernatural, theological and, in this context, it is understandable that a church is established. From our perspective, the incarnation of an OEMMIIWOA (Christ) is part of an explicable scientific-biological context, when one has a holistic view of the WAAM WAAM (cosmos). That in biological evolution, arises an OEMMIIWOA is logical and necessary as that a rock is attracted by a star due to gravity.

This is why, for a religious spirit on Earth, the notion of a OEMMIWOA leave him cold, disenchanted and, perhaps, disappointed, among other things, because the image of WOA that we accept has nothing to do with the theological notion that many religions have shaped on Earth around an anthropomorphic type on the mental plane, paternal, which punishes and rewards, super-smart and creative.

For us, all those ideas are in the realm of myth, which is due to the historical evolution of your earthly social network. Based on a true story, interpreted as miraculous nature, has developed a doctrinal treatise that gave shape to a new religion, Christianity, and the building of various churches as distinct interpretations, that distorted the message of this OEMMIWOA.

Our cosmological conception is based on sound science. We know that we are immersed in a WAAM-WAAM (bicosmos) and information flows that make possible all its wealth configurative proceed from two centers or poles. One of them is the information generator by antonomasia (resonance). All possible configurations of matter, all the possibilities for BEING, i.e. all terms that you might conceive of the existence perceptible and imperceptible by the senses and other sensitive organs imaginable, have their origin in this pole.

All imaginable forms are, however, not possible for all real beings. For example, our brain can imagine OEMII or humans, of a size of one millimeter, but as a biological entity would not be possible. Remember that a linear reduction of 1/103 (a thousandth) would result in the volume of internal organs to 1/109. Reduction of biochemical metabolism would be proportional to the mass. On the other hand, we observe that chemical molecules could not be reduced in the same proportion, so that a cell of this "hypothetical" man couldn't harbor complex architecture which occurs in our cells. For similar reasons, it would be inconceivable to have bugs on OYAGAA with sizes up to dozens of meters, or a star composed exclusively of protein chains. Possible forms of being must be consistent with the body of bio-physical laws that govern the WAAM-WAAM (the bicosmos). This pole or cosmic matrix information shall enable, by generator transfers, entire configuration of universe multiples. Without its existence, the cosmos would be like a gigantic crystal isotropic configuration, amorphous, lacking configurations or singularities and therefore lacking information. (The term crystal, we use not synonymously with geometric architecture of ordered atoms that would not be isotropic, but to describe an endless IBOZSOO-UU chain in complete disorder, wherein the transmission of light cannot be done and with infinite entropy).

The coder cosmogonic center of these possible configurations, we call WOA. WOA coexists with AIODII, i.e., with the formed reality. One configures the other model [manufacturing]. But it is not important that you identify our version to any conception of God. A quick examination of the two concepts can accept this parallelism. WOA = Generator tantamount to God = Creator, as theologians your design.

But the image of God is quite different in the context of Earth religions that present God as an anthropomorphic being, infinite goodness, thinking being par excellence, perfect, and father of his creatures. Moreover, the supreme existence appears to have been revealed to prophets in your religious and pious context.

It will interest you to know that our idea of WOA was inferred in a scientific way, and not in a theological way. It is true that its existence was proclaimed by UMMOWOA in a historic time period when the science could not have been known to him. But for us, a “revealed” concept is lacking probationary value. The society of UMMO is not sensitive emotionally in the same way as the social network of Earth. A religion, in the sense you grant to this term (a Union of a human being to his God, with good faith and obedience to the laws and doctrines) could not ever take hold and grow on UMMO. For us, we do not see WOA as a father, nor conceive that we can meet or accept any concept by way of faith alone. Only reason and scientific evidence is used to build the body of our doctrine. (Note that in all our letters we have been faithful to this principle, as our brothers have always stressed that you do not have to believe us on our word alone that we are OEMMII travelers from UMMO. This is because we assume that you should not, as we do not, believe anything in absolute terms that is not proven.)

If we manage to accept in time the word of our UMMOWOA, it is because the evolution of our cosmogony allows us to accept the reliability of the origin of the doctrine. UMMOWOA, just like Jesus, did not claim to establish a Church or a religion.

The difference between what occurred on our two different planets is that our UMMOWOA was born in an advanced society where the rigorous keeping of history did not allow for the creation of myths, and the Jesus of Earth was born in an era in which the language was metaphoric, the science did not yet exist, and the dominant ideas were irrational and strongly influenced by magical conceptions. Thus, this explains why his message was altered, although essential elements have survived, and why his historical figure suffers enormous distortions.

To understand all this, we have to continue to give nuances our words in regards to the concept of WOA. WOA is the source of cosmic information. However, information lacks meaning without the support of material or a carrier of energy (a medium). The symbiosis between WOA and WAAM is of infinite mass. WOA transforms information into reality in the heart of this WAAM.

WOA also establishes a relationship of attunement with these structures we call the human brain, but only in very special circumstances. What is the scientific basis for this relationship? You must understand that in the context of these pages of disclosure, it is almost impossible to explain the very complex mathematical model on which this is based. We will use a metaphor or an image. WOA resonates with the WAAM BUAWA BIAEII (universe which codifies all the information), for an effect similar to what you know in physics as the resonance. If you place at a distance two violin strings and make one vibrate, the propagation of pressure waves on the second interacts and oscillates (self-induction). A similar effect occurs between two circuits provided with an inductance and a capacitance.

The universal center of pure information represents WOA. You can imagine it as a gigantic archive where you could find, mathematically codified, the configuration of a vegetable plant, the resolution of a system of differential equations, the structure of a building, a machine to generate coherent light (a laser), and, we repeat, any entity or any possible thing in the multiple universe.

Eternally, WOA coexists with the WAAM B.B. as the real brain of the multicosmos, which he (it) modulates with this particular resonance effect. But we observe in reality, it does not transfer all the information at once. The multicosmos is like a big cybernetic organism, which corrects itself.

Within the different universes, thanks to a neguentropic current, life is born based on the biomolecular complexity available on a multitude of cold celestial bodies (remember that a cold celestial body does not necessarily have a global origin, but sometimes results from old stars which cooled and keep a certain level of internal heat). The living organisms continue to perfect in complexity; the density of information in the space of structures accumulates. It is about nervous systems.

The outcome of this complexity is the human brain. Its architecture is complex enough that it achieved a qualitative leap, putting itself in touch with its BUAWWA, and it enriches to the maximum level of its connection to the B.B. (BUAWAA BIAEII) or collective consciousness, the experience of the grand brain of the universe, the WAAM B.B.

Note that this grand system whose architecture is built by the BUAWA B.B. (the global brain) suddenly becomes conscious of the universe that surrounds it. It is like a sensor that captures B.B. configurations, namely the cosmos, galaxies, stars, mountains, animals, rocks, and elaborate artifacts. The B.B. inquires and informs itself of its own development. It is as if the multicosmos was a gigantic being whose brain and hands are the WAAM B.B. This intelligence shapes the clay material in different worlds, concentrating in the form of atoms, star clouds, planets, mountains, and alive beings. However, to shape, it must "see." Its eyes are the various brains. They transmit information to the B.B., and in turn, it corrects the deficiencies of the system with templates provided by WOA. The "hands" are nothing but the physical, inter-universal influence of the imaginary mass that propagates from one cosmos to another by producing folds of space, and thus the configurations of mass and energy modulated by the information of the B.B.

See how, in WOA, the information center is static, whereas in the multiple universes, the WAAM B.B. is in resonance with WOA, and the information is dynamic. Because of this difference, we warn you that the comparison with two violin strings is only didactic and metaphorical, because in them, the effect of echo shows itself by simultaneous dynamics. We used the verb OYAGAA, "to generate" as representative of the action of WOA, because he is more familiar to you and reminds you in a didactic way of the concept of the verb "to create", so dear to the theologians of Earth. WOA is not the being you imagine as God: paternal, intelligent, thinking, with an anthropomorphic structure, who decided to create a universe and to put in it, creatures in his image that he will reward after their death, if they succeeded in carrying out his laws. WOA has nothing to do with this mythical being created by the spirits of the men of Earth. Here the verb "generate" could be translated as "represent" by a special effect of cosmic resonance. The models of information are transferred to the WAAM B.B. to facilitate the evolution of time of the configuration of a network of the universe. A simple analogy could help familiarize you with our cosmology.

The WAAM B.B. is like the brain of a potter whose tired eyes (the brains of OEMMII) contemplate a lump of clay (matter and energy). His hands (the imaginary mass whose "tentacles cross the borders of the separate universes) will shape an amphora. But to do so, two mental processes are necessary: first, an inspiration to draw a design, the information model that represents a container. For this, he looks at an old book of pottery (WOA), which subtly suggests its form should have the shape of an amphora, but above all, he must learn to correct the shape as he works, manipulating it with his hands, watching as it takes shape, becoming aware of the difficulties entailed in handling a viscous substance.

When we give you this word "generate" in our writing to the people of Earth, we do not make a reference to this hypothesis of yours of a divine function "to create some material out of nothing" but rather the concept of engendering "IMAGES of IDEAL BEINGS" in the WAAM B.B., which is responsible for invigorating or shaping the universe and coexisting with WOA. Things have not been created, in the sense that you give this word, by God.

We have a profound respect for your religious conceptions of the entities you call Allah, God, Jehovah, Brahma... But as you have just seen, our concept of WOA has nothing to do with your theological ideas. You do not have to feel forced to embrace our idea of WOA, which for us is a scientific concept, but which for you is nothing but an idea on pieces of paper whose origin is obscure. For this, each of you has to continue to be faithful to your traditional beliefs as we have always suggested to you, and to read our reports as if studying the customs of some exotic and distant tribal village.

(Excerpts from letter D792-1)

Spatio-temporal Man

The cosmic physical reality rightly requires a design for man, not as a three-dimensional being, but as a tetra-dimensional being, at minimum. The cosmic ties between the OEMMI, their planetary B.B., and their BUAWA are so inseparable that implies to include the Man in its temporal dimension.

We must not speak of Man Homo Sapiens, but the spatiotemporal Man, Homo Sapiens spatiotemporal...

This distinction is very important and is not merely a conceptual philosophical vision, but a cosmo-physical reality.

Spatiotemporal Man permeates axes of space-time, leaving a trail etched as car headlights at night on a film...

The trail engraved by OEMMII in space-time is AIOOYA AMMIE along the axis of time. It is no longer perceived by our eyes, only mnemonic archiving allows us to remember that we have a past. However, the spatio-temporal trail engraved by OEMMII still exists in cosmic IBODSOO. The spatiotemporal Man is a long snake with the human body at its head. Like the snake, his body and head are inseparable.

Man-spatiotemporal is under the control of its BUAWA which determines and directs the guideline of the life of the OEMMII. The neural brain allows humans to adapt and implement the guidelines of its BUAWA.

D41: In the space-time continuum (as incorrectly named by EARTH physicists), the human body is a "fold" over SPACE (depression through a fourth dimension) we can define mathematically ten dimensions. In fact, MASS with a Volume and associated time. We cannot conceive the time if it is separated from the other magnitudes.

People who have a poor scientific training consider the man as a three-dimensional (volume) being, living various facts in the flow of time. For him, there is only the memory of past events. The only reality is the present, and the future does not exist yet... This description of the world is absurd and childish.

Imagine that are arranged along an axis that represents the dimension TIME various situations (EVENTS) that has lived, lives, will live a man throughout his LIFE.

Space and time are so closely associated that if we joint into a single graphical expression, a single image, all these situations or events that saw the man throughout his life, we get a strange BEING with four dimensions (volume + time) that looks like a huge OEBUMAEI (sort of long "donut" or "pudding" popular in the region AADAAADA on UMMO), the section will represent a man if we were cutting it into slices.

UMMO cosmologists call this tetra-dimensional being: OEBUMAEOEMII

WOA grants the soul a prerogative that is transcendent.

IT MAY CHANGE ONCE AND FOR ALL THE FORM OF OEBUMAOEMII (MAN-PHYSICAL/SPACE-TIME).

This means that if WOA (GENERATOR or GOD) generates and creates the physical body, in establishing the characteristics of physiology, he concedes the BUAWA the ability to shape the conduct of the body throughout time, once and for all.

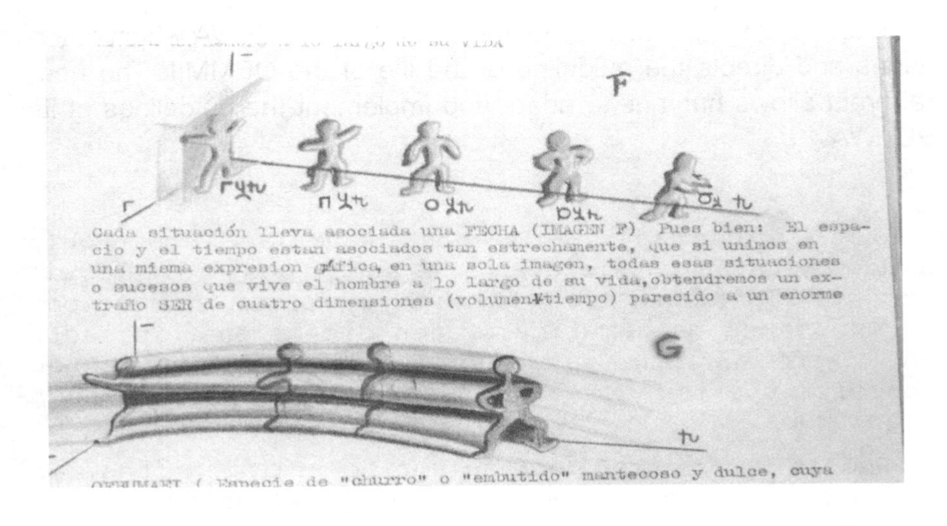

Summary and Conclusion

The science of the twentieth century based on rationalism, positivism arrived at the dawn of the 3rd millennium to its limits. A new cosmological paradigm generating a new framework for scientific thinking is needed to continue to be able to explain complex phenomena around us.

IBOSDSOO universal substrate, similar to the concept of "String" is both simpler and more powerful in its ability to describe physical and mathematical developments. Angular axes OAWOO, near by the concept of vector dimensions, also allow a clearer view of the phenomena across the cosmos.

This new perspective of the cosmos is also extended, one hand by infinite universal dimensions, and by the 10 major dimensions described by Oomomen in particular, which give us the vision of a space, where travel and communications in multi-cosmos are intelligible.

This IBOSDSOO model fits perfectly consistent and in line with the developed or approached theses for millennia in many terrestrial crops, Vedic Maya, the Leibniz "monads", through Fourier and Niels Bohr "quantum interconnection" and in 1952 the "quantum potential field" of the holographic model of David Bohm. This long intellectual journey found one of his decisive evidence in 1982, with Alain Aspect and his team, what is now called "quantum entanglement" and which is non-local.

The switching of angular axes can be complete to move to another part or three-dimensional framework for LEIYO effects of "anti-gravity". Black holes do not really exist; it is simply great mass neutron stars that end their lives into anticosmos.

The anticosmos contains large amounts of anti-matter that affect our current cosmos. We see these masses by its gravitational effects and changes in the rate of expansion of the cosmos that contemporary science understands in terms of "black" matter and energy.

All anticosmos cosmos and are separated by a membrane XOODII that contains masses without volume that carry the gravitational effects from a cosmos into its opposite.

Tetravalent logic allows new mathematical developments and consistent homogeneous transcendental understanding of cosmological entities hitherto relegated to the domain of metaphysics.

The being OEMMIIWOA often referred to as a prophet finds its place in a logic of human evolution. It is physically a mutant human whom directly will capture stream of communication from Meta brain BUAWE BIAEI. Meta-brain is a relay entity of that we call "God", it can try to guide human evolution path towards more intellectual and spiritual.

There is no religious vision in this new paradigm, and I condemn in advance any person attempting to hijack this work in a religious or sectarian purpose.

Instead, this new cosmological vision reconciles Science and Theology in a new unified paradigm of rational-positivist approach including Metaphysics.

*

THE LIFE EMERGENCE

All these assumptions about the genesis and evolution of the entities of the universe will allow us to describe the key points and the foundations for the emergence of living. Emergence of living cannot be understood without an overview of the different cosmological entities.

D731: There are as many BB as AYUUBAAYII (networks of planetary alive beings) throughout the WAAM-WAAM. There is a correspondence between each set of alive beings on a planet and its corresponding BB.

First, we must try to understand how the content of the modeler BB-global is created. Regardless conventional astrophysical mechanisms what happens, then, when stars are created?

Then we can ask ourselves about the mechanisms that led from the star "driver" to the emergence of Life itself. This is my own theories developed on the basis of my interpretation of the Ummo cosmology.

THE CONTEXT OF A META-BRAIN PLANETARY-BB

The Ummo people tell us that the parameters of cold stars are known by BB and that given, these parameters, life is possible or not. So we know that:

- planets are associated with their BB. For all beings by an intracellular communication channel, BAAYIODUU. For humans, by a cerebral communication channel, the OEMBUAW.

- the link between the BB and its planet is gravitational.
- the gravitational link also transmits the planet's parameters to BB.
- BB is in the WAAM-UU, the planet's gravitational link to BB passes through the channel of the intercosmos layer XOODII by a border effect.

The hypothesis of "Krypton constant"

The stars are governed by physical laws. These laws include physical interactions between the entities of the cosmos WAAM and the WAAM-UU transiting through Imaginary Masses in the layer XOODII WAAM. In some cases, the celestial bodies could be linked with a meta-brain, a planetary-BB.

The assumption is that the celestial bodies with water in the liquid phase and a certain concentration of krypton, and only these planets, are potentially able to connect with a meta-brain, a planetary-BB.

The krypton atoms must be dissolved in water, and the concentration of krypton in the aqueous solution is such that it allows being bound gravitationally to krypton atoms.

When the configuration of krypton atoms becomes "ideal", then it resonates with the gravitational frequency of the planet, and then the connection with its planetary-BB location. This krypton atoms configuration is a kind of archaic BAAYIODUU, that is to say, it is not yet part of a alive being, since there has not yet...

The krypton concentration of an aqueous solution of these celestial bodies is in such a range that enables it to bind gravitationally krypton atoms in a chain. This chain is structured in an "ideal" configuration, which resonates with the gravitational frequency of the celestial body.

This relationship between the concentration of krypton in aqueous solution and the gravity of the celestial body allowing the ' ideal ' configuration is probably corresponding to a specific threshold, I will refer to as "Krypton constant "This constant is therefore underlying the phenomenon of resonance of the chain of atoms of krypton.

To avoid confusion, note that the resonance is produced following variable data. It is obtained by matching:

- a H2O mass variable
- variable gravitational frequency of the planet
- with a concentration of krypton which is also variable.

If we were to try to make such a constant, simply, it would be the result of a ratio of the order of variables, for example:

Cte Kr = [Mass (H2 O)/Mass (Kr)]/Frequency (G) ...

The adequation of these parameters: the amount of water on the planet, its gravitational frequency, the concentration of krypton in aqueous solution, thus causing a specific gravitational resonance. This specific gravitational resonance causes a border effect LEIYO allowing codification uniquely with the planet. In summary, this informational link planet-krypton-BB is unique, gravitational and produced through a border effect channel. This is the archaic BAAYIODUU, that is to say, it is not yet part of an alive being, because there is none.

This border effect generates a connection to a planetary-BB initially empty in the WAAM-UU. This first connection initializes the BB with the structuring parameters of the planet (its mass, geological, etc.). Communication links planet-krypton-BB is then established, it dynamically informs the planetary-BB on the status, the evolution of the parameters of the planet.

As long as the planet emits gravitational waves that resonate corresponding to its krypton density, this link identifies the celestial body in the planetary-BB. Krypton in the aqueous solution became receiver of gravitational frequencies emitted by BB transmitting scalable templates.

All these interactions pass through Imaginary Masses into the layer XOODII WAAM.

Diagram of the link planet-krypton-planetary-BB

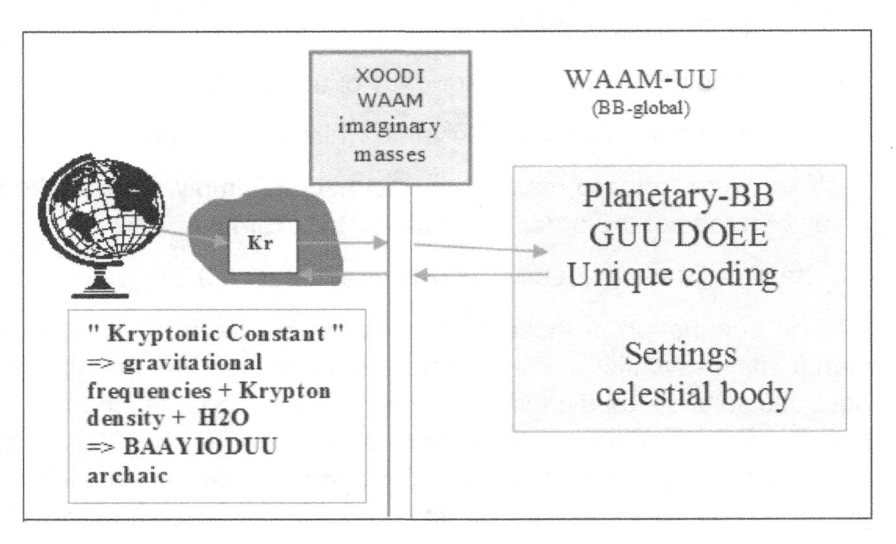

Figure 1-p1 :

Schéma du lien Astre-krypton-BB suivant une 'constante Kryptonique' sous-jacente correspondant au seuil de concentration de Krypton en solution aqueuse qui permet la création du BAAYIODUU qui lui-même entre en résonance avec les fréquences gravitationnelles de l'astre.

DETAIL OF THE HYPOTHESIS OF "KRYPTON CONSTANT"

Idea of the emergence of life from inert matter assumes that it changes state and acquire new properties, those of living.

Main property of the alive beings is the ability to self-reproduction, the second ability is evolution-adaptation. A combination of these two properties is an emergence that does not have the inert material. And as we have seen entropic beings, inert, lose information, while neguentropic beings alive, and absorb information from the external environment.

In other words, a being living is self-replicating and its copy has the ability to evolve, based on information acquired from the external environment.

General phenomena of emergence have seen multiple approaches: holistic, vitalistic, reductionist or emergentists revolving around the subject by illuminating their various points of view, but none of these approaches was able to describe the essence of the emergence of living.

To explore my hypothesis, I will remember for initial principle of emergence, the Hegelian philosophical approach and its mathematical and systemic equivalent described by the Non Linear Dynamic System model (NLDS)

In these explanations, I propose to define the emergence as follows:

There is emergence where a system in a stable condition with original properties, bifurcates at a critical point to one or more stable states with new properties. The bifurcation maybe a discontinuity.

The quality of emergence, that is to say the level of its transcendence, depends on the density of information by interrelation elements composing the original system and the network architecture of the system.

The semantic analysis of the term EIDUAYUUEE gives us an idea of how Ummo people perceive this concept close to the "emergence"::

"EIDOAYUEE is the obvious fact to you that a network has properties and performs functions that do not have the elements that compose it."

The translation of EIDUAYUUEE is "The concept identifies an event that depends on the network architecture."

GENERAL INFORMATION ON NONLINEAR DYNAMIC SYSTEMS

When a control parameter P given reached the critical threshold Psc, a stable system S1 branches to 2 or more possible stable states S2, S3, Sn, Sn +1.

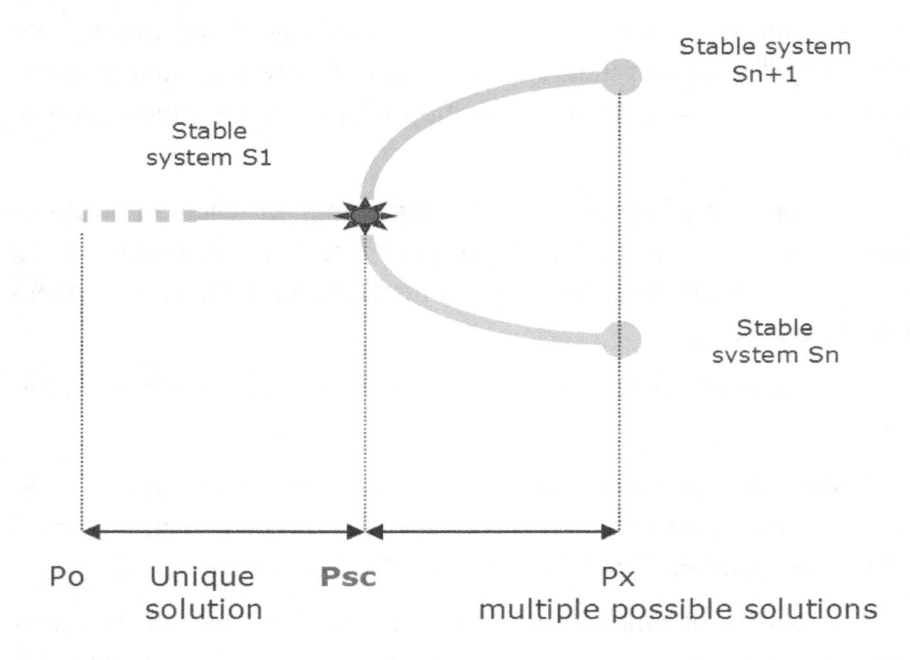

NONLINEAR DYNAMIC SYSTEMS FOR THE "KRYPTON CONSTANT"

Suppose a given environment whose structural parameters (mass, gravity, etc.) are fixed and sufficient. The structural parameters are NLDS components whose values are necessary and must also be applied has sufficient to produce a system state change. The structural parameters not produce by themselves the change of NLDS state.

In our case of applications the control parameter is the "Krypton constant". Remember that it would result from a ratio like: Cte KR = [mass (H2 O)/Mass (KR)]/Frequency (G) ...

We can simplify and reduce to P = "Kr concentration" for a medium whose structural parameters (mass, gravity, etc.) are fixed and sufficient.

The initial stable system S1 is composed of atoms of krypton in aqueous solution.

The bifurcation occurs at the threshold of the krypton constant, which correspond to a gravitational resonance effect between krypton atoms in aqueous solution and gravitational frequency of the planet.

This would produce the organization of the krypton atoms chain in a stable system named BAAYIODUU-archaic and has a connection effect LEIYO to the planetary-BB.

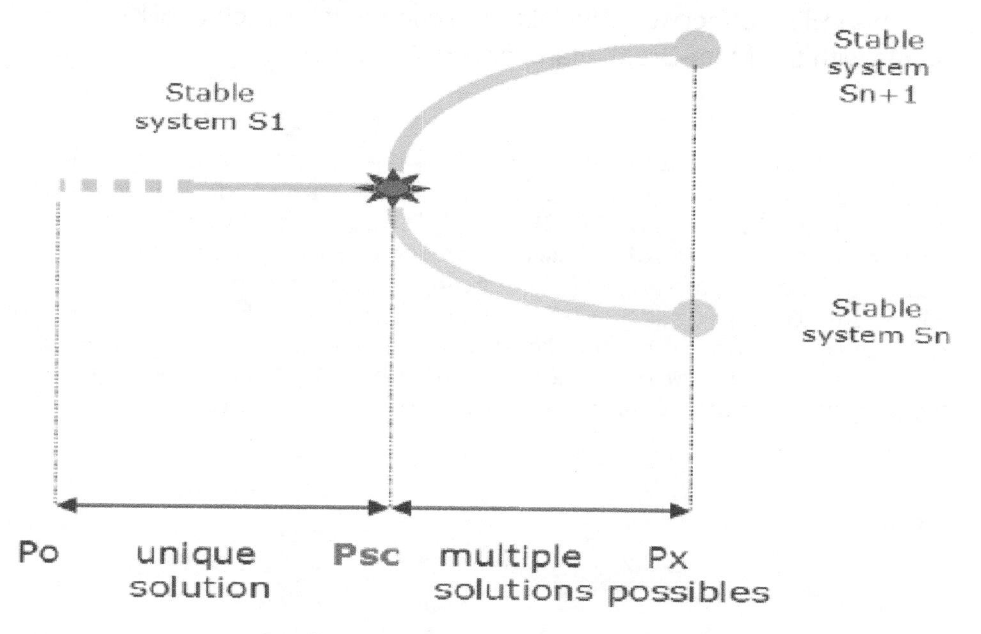

gravitational resonance with the Krypton of the planet

Point LEIYO

<u>Entropic state</u>
Kr atoms aqueous solution

<u>neguentropic state</u>
BAAYIODUU-archaïc string of Kr atoms

THE LEIYO EFFECT AND KRYPTON

The “lei”—“illo” concept describes “a set of cosmological phenomena” trivially translated by “effect border “, that occur in what Oomomen call XOODI WAAM, that is to say, a “layer” between two cosmos. The “true LEIYOO WAAM is a highly complex phenomenon that involves the transformation of an ibozsoo uhu network." The semantic analysis of the term LEIYO expresses "Changing concept identifies a set of entities" that I translate as "isomorphic transposition to a set of ibozsoo uu." Specifically, the LEIYO effect that would be involved in the threshold of the krypton constant is a gravitational phenomenon of "resonance" with the parameters [mass (H2 O)/Mass (KR)]/Frequency (G) ...

This LEIYO effect would initialize the connection structure of Krypton atoms with the Meta-Brain, planetary-BUAWE BIAEI.

D58: "These [Krypton atoms] were at the ends of the helicoid of deoxyribonucleic acid chain by forming several pairs (FIG 58-2F8) (a total of 86 bi-atomic groups) that ran on common orbits and plans orbital, substantially parallel, enjoyed a common axis (the axis AB in Figure 58-2F8). This axis at the same time described a harmonic vibratory motion whose frequency and amplitude were based on the TEMPERATURE (0.2 Megacycles for a temperature of 35 ° Earth centigrade)."

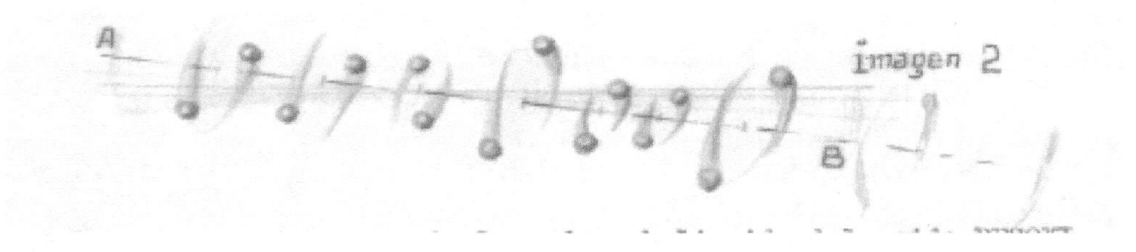

D731: "How is it possible that the electrons of a krypton atom behave in the B.I. [B.I. BAA = IYODUHU (union factor between BB and chromosomes] and the O. [O. = OEEMBUUAW (krypton factor that unites BB with the brain of an OEMII)] such a special way to work? These are the subparticles M imaginary, "the other side" of the border, take this action."

When alive beings with DNA will be formed, this means that there would be 43 Kr atoms thereof to each end of a chromosome, stored in each telomere. Each pair has a stable Kr2 orbital plane. All these Kr2 pairs would have orbital planes parallel to each other. In the present state of our hypothesis alive beings are not yet formed.

At the critical threshold of krypton concentration in aqueous solution, gravitational frequencies of the planet and krypton resonate, it is LEIYO effect. This generates the chain of pairs of atoms of krypton and they synchronize their orbits. Their electron layers initialize the planetary-BB incoming connection. This first connection initializes the BB-planetary parameters of the planet (its mass, geological, etc.). The link planet-krypton-BB communication is then established, it dynamically informs the planetary-BB.

So, it would be formed a BAAYIODUU-archaic which will be integrated into alive beings.

Summary and Conclusion

We saw in the assumption that, given an universal constant called "Krypton constant," the planet that contains a certain concentration of krypton in aqueous solution generates a particular assembly of krypton atoms that can connect and initialize a planetary-BB. So, it's an archaic-BAAYIODUU which is not yet part of an alive being, since there has not yet...

Still following this thread, so I will continue to develop this hypothesis in a consistent manner for the emergence of Life with quite numerous and precise information we have on the link BAAYIODUU of alive beings with the planetary-BB.

With Ummo documents guidance, I will continue the discussion with a complementary hypothesis that develops explanations of the gravitational flow linking alive species to BB through the krypton.

*

The oriented evolution of Life

We know the phenomena of evolution by Darwinian natural selection or rather neo-Darwinian as described by Stephen Jay Gould. We may think that solar activity and the proximity of the Earth from the sun generate very high rates of mutations that make Darwinian evolution very visible on Earth. This mechanism would have overshadowed another mechanism, slower and therefore less visible.

However, when we see a beautiful orchid flower turn to mimic the smell and visual appearance of a female insect in perfect synchronism with the coevolution of the male insect development, we can ask how such synchronization is possible? Darwinian approaches lack of answers...

Notions of "morphogenetic field" developed in the 1920s by Hans Spemann, Alexander Gurwitsch and Paul Weiss, then in 1981 by biologist Rupert Sheldrake. These "morphogenetic fields" would be decisive in the behavior of alive beings inheriting habits of the species by "morphic resonance". Rupert Sheldrake thought morphogenetic characteristics are structures of "probability" in which the influence of the most common kinds of the past combine to increase the likelihood that these types reappear.

Rupert Sheldrake therefore made an intellectual improvement and therefore senses the problem of the creative potentialities of the Real Absolute, the AllODll, without perceiving the profound cosmological dimension or even the physical foundations.

Yet long before him the psychiatrist Carl Jung understood that there is an informative exogenous structure for humans, which provides the "archetypes" at a deep level, a collective unconscious shared by the entire human species. Unfortunately, as any biologist, Rupert Sheldrake will not know see the scope, nor the connection with his own research...

We saw in a previous hypothesis that archaic-BAAYIODUU is dissolved in water. Given that the context is the following:

- The planet must have water in the liquid phase.
- On Earth, in water were amino acids into the famous "prebiotic soup" more or less enriched by panspermia.
- As Oomomen explained, water is an "informational amplifier" for exchanging information captured by the Krypton until within alive cells.
- There is a mechanism for exchanging intracellular information with the planetary-BB, which led to put under control of genetic mutations, i.e., placed under control combinations of amino acids together, to modify the genomic system.
- planetary-BB and the planet are interacting, and they follow the universal laws of phylogeny and orthogenic.

Phylogeny and orthogenic

D 792: "a law unknown to you and what we call BAAYIOODISXAA (Cosmic Balance biological) and commonly referred to you as species evolution, process governed by rules emanating from BB (orthogenesis) through IDUGOOO (Changes or successive genetic mutations)."

D57-3: "The formula that expresses the BAAYIODIXAA UUDIII is a complex function in which are embedded in a variety of settings ... formula that expresses the conditions for biological balance measured in a given environment."

The word BAAYIODIXAA UUDII can be translated as: "dynamic interconnection evolution of a set of evolution parameters whose morphology depends on (for alive beings)"

This function expresses the mutations control of alive beings by BB via BAAYIODUU. Mutations under the control of BB then follow evolutionary patterns, where each phylum probably corresponds to an emerging class, a model or pattern into BB.

What can happen between archaic-BAAYIODUU and amino acids? How can it emerge alive from inert matter?

The Emergence of organic protomolecules

The hypothesis of the oriented emergence of Life is that information from the planetary-BB and captured by the archaic-BAAYIODUU would induce, "catalyze", the amino acid groups to form a self-replicating structure. This self-replicating entity has a primary genomic structure consisting of amino acids that is to say "organic protomolecules". It transmits and receives information to planetary-BB.

Building of amino acids would be done by oriented manner via the archaic-BAAYIODUU. Chemical assemblies of amine functions are "catalyzed", directed by planetary-BB in the range of all "model types" of possible combinations corresponding to the parameters of the planet, thus with a limited number of possible combinations. So it would appear an archaic-RNA.

The replication mechanisms of archaic-RNA is probably due to ribosomes. This is a purely physico-chemical mechanism. Only internal mutations in the sequences of archaic-RNA are under the control of planetary-BB via archaic-BAAYIODUU.

It is an oriented emergence by planetary-BB via the archaic-BAAYIODUU.

The self-replicating combinations archaic-RNA-based appear in a choice of "standard models" after the classical orthogenesis evolutionary principle. The self-replicating entities are then made according to the optimal model, among other possible models in the planetary-BB. The information processes between krypton and BB continue in the same manner. The informative content transmitted to BB evolves qualitatively with the complexity of the entity.

If the structural parameters of the planet allow, for reasons of stability and reliability of structures, the archaic-RNA is probably quickly integrated into a membrane structure protecting the whole structure.

It is not explicitly stated in the Ummo documents that the first alive self-replicating unit is a structure with a membrane type like coacervates, for example. Nevertheless, the following of evolution is based on self-replicating entities membrane structures.

Figures 1 and 2: the oriented effect on the archaic-BAAYIODUU in aqueous solution within the amine functions allows the building of an archaic-RNA according planetary-BB models.

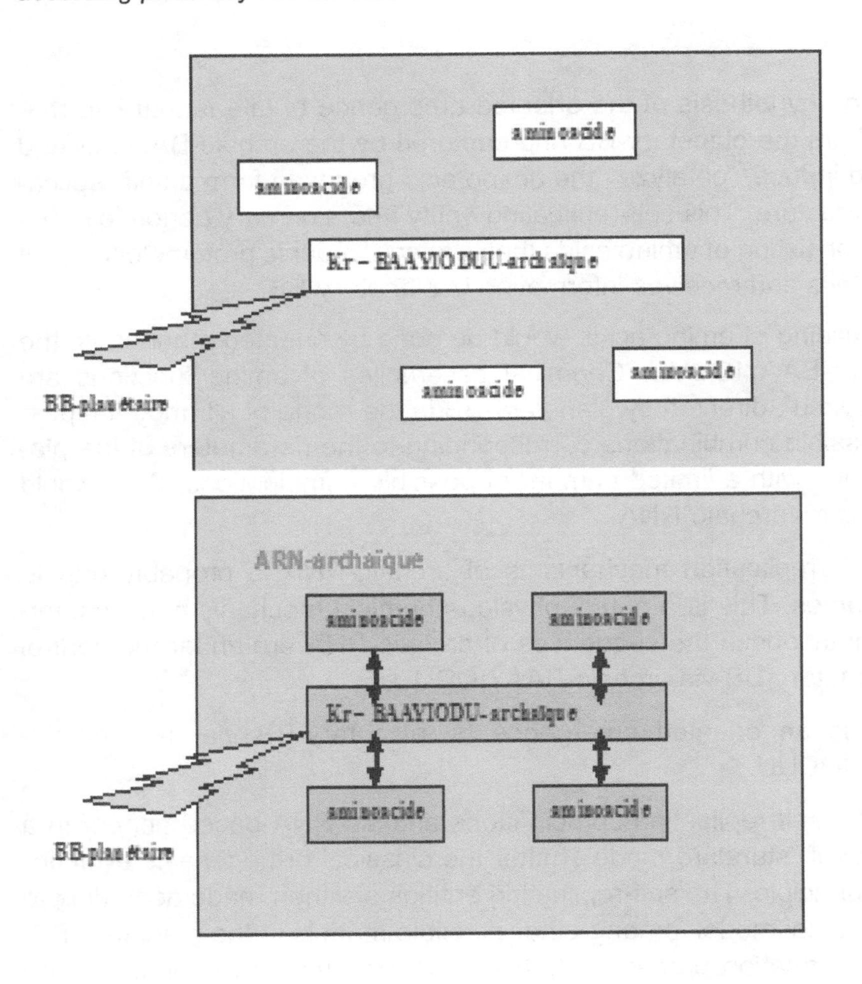

Figure 3: The archaic-RNAs is incorporated into a membrane structure and provide the first self-replicating entities, negentropic alive beings.

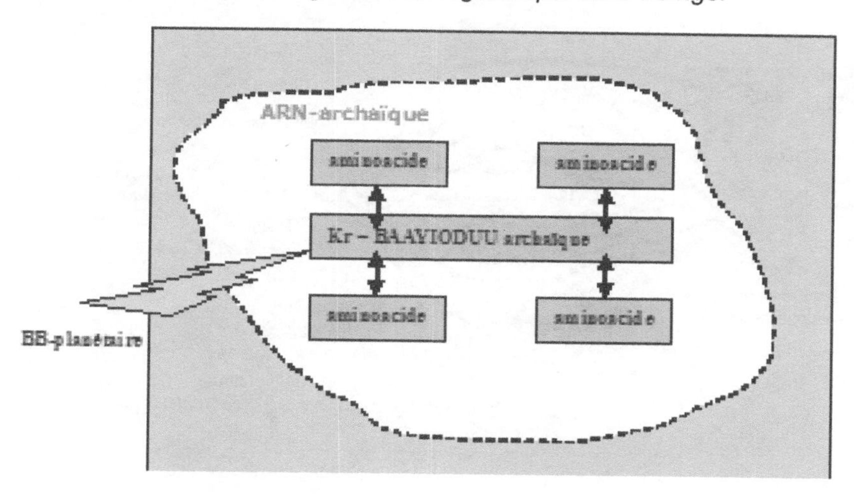

SEMANTIC ANALYSIS OF THE TERM UUDIE

Although it is not stated verbatim, we can see graphically the Ummo scheme, an arrow indicating the influence of BB on these "proto-organic molecules" and "primigenes bodies" which are also the first alive beings self-replicating.

The translation of UUDIE:

- the dynamic dependence has a form that identifies a concept [sent to or received from BB].
- faculty of perception.
- "biosensor".

The translation of UUDIE BIEE:

The biosensor has a connection that identifies patterns [into BB]

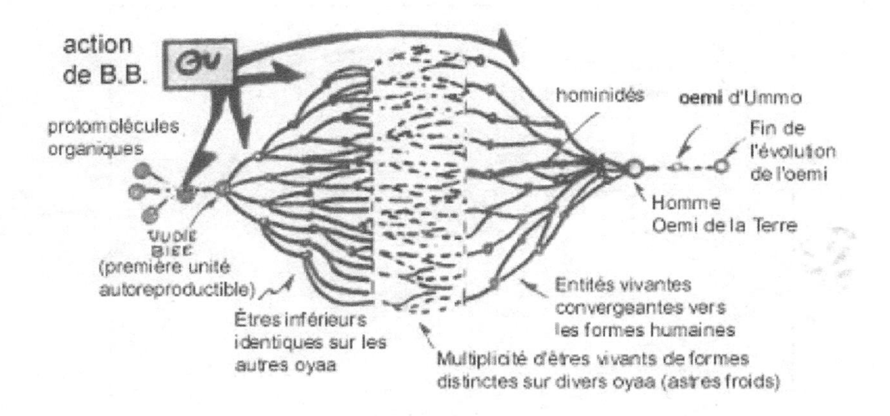

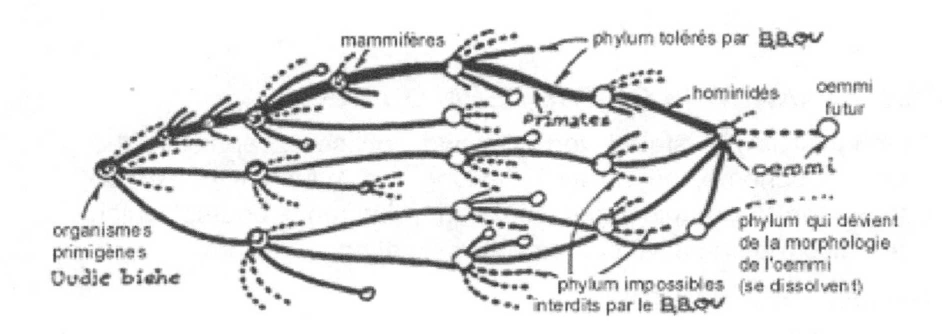

SUMMARY OF EVOLUTION PHASES: RNA, PROTEINS, DNA

Prebiotic soup, due to various phenomena of panspermia and peptide synthesis, contains krypton and amino acids in aqueous solution.

The constitution of archaic-BAAYIODUU is performed by the first LEIYO effect threshold of “krypton constant”. There are connection and initialization of the planetary-BB. This is the beginning of the scope of the concept of Evolution BAAYIODIXAA.

The archaic-RNA constitution is done by the assembly of amino acids under the action of the archaic-BAAYIODUU in aqueous solution. That is, the appearance of the first self-replicating alive beings.

Appearance of protein.

DNA appearance, probably due to the activity and the development of RNA viruses.

Conclusion on the Life emergence

What marks the difference between the inert and the living is that the inert material is subjected to physico-chemical laws, while for Living the physico-chemical hazard is controlled and "directed" by models into the planetary-BB. When crossing the threshold of the "krypton constant" inert pass under the control of the planetary-BB and architecture of alive structures go up. The archaic-BAAYIODUU is consisted and allows the assembly of amino acids archaic-RNA that is replicated by physico-chemical mechanisms and quickly encapsulated in membranes.

The first living entities archaic-RNA based are under the influence phylogeny and orthogenic laws which are implemented according to genotypic and phenotypic patterns contained in the planetary-BB which controls mutations.

The laws of evolution expressed by BAAYIODIXAA concept follow a more rapid evolution than would be the only statistical random, quantum laws governing microphysical phenomena. Thus, Darwin's finches had mutated through viable genotypic and phenotypic patterns, much faster than if the changes were due to chance alone. Darwinian natural selection does the rest...

Figure: The controlled emergence of Life

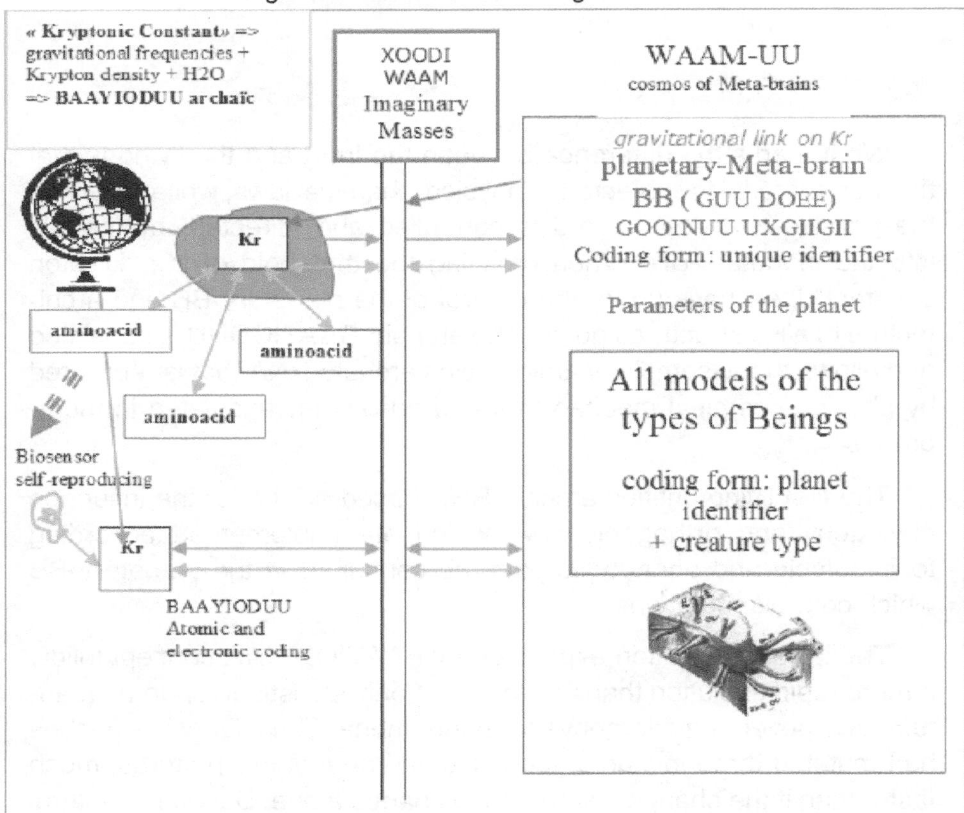

*

INFORMATION FLOWS OF ALIVE SPECIES

In line with previous assumptions, we will see a hypothesis describing the gravitational flux linking alive species to BB through the krypton. On this subject we have a few elements to begin.

We know from the oomomen's writings that:

- BB contains all the "standard model" of alive beings on the planet
- BB transmits information to alive beings that validate the evolution of beings in accordance with "standard model"
- The link between BB and genomic system of beings through resonance with the Krypton, is of gravitational type
- The BAAYIODUU link of the genomic system (all DNAs, RNAs, etc.) with BB makes a relationship between an alive being with all the "standard models" of possible beings of the planet.
- All possible genome configurations of alive beings are identified by a unique code. This unique coding consists of all possible configurations of each electron, in the eight sub-layers of each Krypton atoms.

Moreover, we know that human beings' "soul" is in connection with the first cell of the alive being at the time of the genomic fusion.

How the information in the cells of alive beings can they are sent to the right planetary-BB?

How planetary-BB communications can they be received by the right recipients?

INFORMATION FLOWS OF ALIVE SPECIES

As we mentioned earlier, we will assume that for alive beings, it is at the time of genomic replication, RNA or DNA system type, which BAAYIODUU connects to its planetary-BB, and that alive beings begin to communicate with the BB.

Gravitational frequencies emitted by the biochemical processes of alive beings resonate within intracellular krypton causing LEIYO frontier effect that allows the transmission of information to planetary-BB.

If it is perhaps not necessary to know the exact BB source for each individual of information sent to it, it must nevertheless know at least:

- the emitting species corresponding to a template class
- the planet transmitter

This underlies the concept of species is not related to genetic-evolutionary constraints that are only resulting from an upstream process, but ultimately strictly to planetary-BB categorizations.

Each species would correspond to a design type in planetary-BB and each model to a particular level of emergence.

THE IDENTIFICATION OF THE EMITTING SPECIE

Since each planetary BUUAWE BIAEEI is linked to a planet and contains:

- Information of perceptions and mental processes of superior beings (OEMMII)
- Universal symbols, main-ideas of superior beings
- Emotional patterns gregarious superior beings
- Biological information of ecological environment of all alive beings

- Patterns of life forms of all alive beings

H8

—In theory, if a hybrid of two humanities could exist,

to which collective subconscious this individual belongs?

U— to the one he would have more genes.

BB communicates through imaginary masses

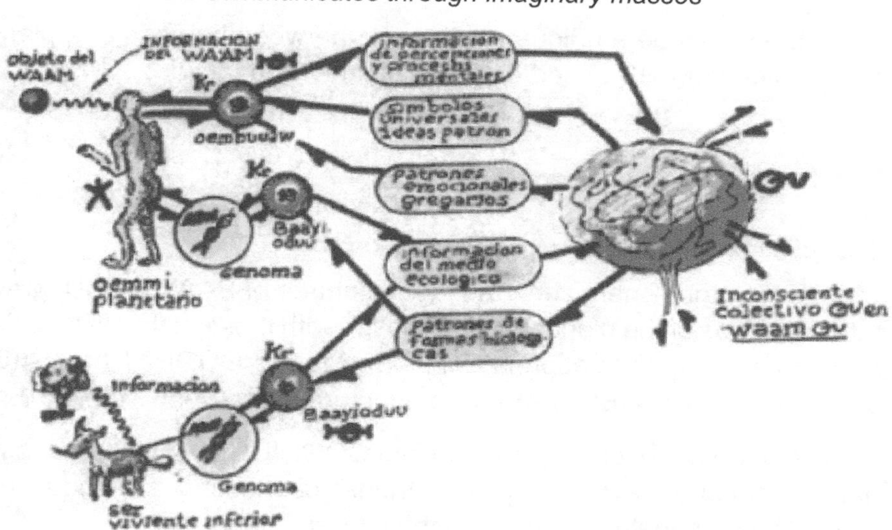

Biological patterns of alive beings are fully encoded in every cell of every alive being, so they can be identified from within the cell.

The gregarious behavioral profiles are not contained within the cell. They need to be identified in the planetary-BB specifically for each human species of the cosmos.

It necessarily has a unique link between human specie and a type of gregarious behavior profile. The assumption is that the identifier of the type of gregarious behavior profile is a gravitational frequency characteristic of the type of the human specie.

The same applies to the connection flow between humans and data from planetary-BB, they must have a unique identifier.

THE IDENTIFICATION OF THE PLANET TRANSMITTER

The identification of the planet transmitter is necessary to ensure the information transmitted by the alive being goes into the right planetary-BB, and not into another!

This implies that gravitational frequencies that identify the planet are juxtaposed to the gravitational emission of alive beings. There would therefore have a major gravitational frequency of the planet in that play the role of a "carrier" and a set of secondary gravitational frequencies for each alive species which act as "modulated frequencies" or "harmonics".

There would be a unique relationship between humans of a planet and its planetary-BB based on gravitational frequency.

THE FLOW OF INFORMATION ENTERING INTO BB

Into Krypton "Energizers" atomic structures of BAAYIODUU, gravitational waves of the planet are juxtaposed with those of the alive being. All gravitational information is transferred to the GOOINUU UXGIIGII structure of BUUAWEE BIAEEI, as quantum energy.

The quantum of energy arriving into GOOINUU UXGIIGII and resonate (stationary waves) with the "strings" connecting the nodules in pairs. They resonate according to their informational type.

Gravitational forces initially be issued by a given specie therefore correspond to information of various kinds, transmitted according to different wavelengths with different content. The GOOINUU UXGIIGII "strings" therefore resonate according to their correspondence to each frequency, then the information is transmitted to the nodules and processed.

THE FLOW OF INFORMATION COMING OUT OF BB

Conversely, GOOINUU UXGIIGII information must be transmitted to the right specie, such as pattern forms of birds must be transmitted by BB to the correct recipient: a bird, not a fish!

BB recorded information into GOOINUU UXGIIGII with initial wavelengths of a given specie. BB continuously emits all pattern forms of all alive beings. Each specie capturing information that is specifically designed. The information resulting from the emission of quantum energy of BB, will resonate with BAAYIODUU configuration with the wavelength corresponding to the concerned specie.

Ultimately, it is the electrons of krypton atoms that receive and "resonate" with gravitational harmonic which correspond to them.

Hypothesis on BB categorization structures

The assumption is that the planetary-BB or more precisely the GOOINUU UXGIIGII, is structured as IBOZOO networks like the evolutionary phyla. Large networks of categorization of alive beings, with subtrees characterizing each species, and other nodes still marking a difference of genotype for the species or the herd scheme, etc.

Some particular nodes including specific species, being characteristic of a level of emergence, variable according to evolution level.

Thus, if several OEMMII phyla exist on a planet, such as Homo Sapiens sapiens and Homo Sapiens Neanderthals do not generate itself two separate planetary-BB, but two separate IBOZOO networks in the same planetary-BB.

This would explain why Oomomen use the same word BUAWEE BIAEEII for several distinct objects:

"D357-2 of 1987: (The confusion you might see just what we call BB (BUAUEE BIAEEII) not only the collective UMMO or Earth Soul, but also the cosmic plan (that is to say the multiuniverse) that contains all BB of different social networks that populate our tetra dimensional universe"

"That is to say, the planetary-BB (the first sense [old] stands for "EESEOMI's COMMUNITY") and the WAAM-UU containing all planetary-BB (our current concept of "COLLECTIVE SPIRIT").

So, 3 different concepts for BB:

- of "EESEOMI's COMMUNITY" the first sense (old)
- of concept planetary-BB "UMMO or Earth collective Soul"
- of concept global-BB (the cosmos named WAAM-UU) "cosmic multiuniverse plan"

Furthermore, we have explicitly in 1966—D33-3:

"You might be objected that perhaps there are several BIAEI BUUAWE associated with different racial groups. We do not believe this view plausible for the simple reason that we NOTE that all human of the Earth are from the same anthropoid phylum."

So there is a single BB for our humanity.

But in the document D1751 we have:

"We invite you to reflect on the contradictions of the Islamic collective soul (BUUAUE BIAEEI)."

The indication "collective soul" characterizes the planetary-BB object.

So that would imply that there are several planetary-BB for our humanity.

To be consistent we should have:

"We invite you to reflect on the contradictions of the Islamic 'EESEOMI's COMMUNITY' (BUUAUE BIAEEI)."

A priori, we can think that this sentence is a serious mistake, as if we use an old language term amid a contemporary sentence. But it is perhaps simply because the term BB is used for different concepts and two OEMMII phyla on a same planet don't generate two separate BB, but two distinct IBOZOO networks inside the same planetary-BB.

This concept would be extremely close to the notion of "community of EESEEOEMMII." This type of concept involving then that there are various IBOZOO networks. This various IBOZOO networks do not mark an emergence specie node, but simply a genotype within the specie and its gregarious scheme...

.

Human brain flow

How the received streams are transmitted, by pairs of krypton in the brain into the OEMBUAW, to the human body?

How are transmitted flows received from BAAYIODUU to the DNA?

How a mutation "order" from planetary-BB is it implemented into the DNA?

As we have seen the flow between the planetary-BB and the electronic layer pairs of Krypton of OEMBUAW from the brain or flow from the BAAYIODUU to cells are LEIYO gravitational effects. So how gravitational flow can be transformed into a biochemical flow in the brain?

D58-5: To summarize we will tell you that the Crab cited, will pick through his eyes light stimulus coloring rocks (BLUE-GREEN). This causes a series of metabolic alterations (i.e. biochemical) stimuli immediately encoded in the form of nervous impulses affect single organs of his embryonic nervous system. In this case it is the levels of Potassium and Nitrogen which are altered so that the cell IS ADVISED of outside conditions, in the OPTICAL field.

Through the Cell Membrane ionic balance transfer is altered and cellular metabolism undergoes a series of changes ranging from cytoplasm to nucleus.

Alterations occur at the most superficial oxygen atoms sublayers that make up the water intra-cytoplasmic molecules automatically producing variations of the electronic gravitational field.

[...] Some oxygen atoms of the water components contained in the cellular cytoplasm, undergo orbital excitation in their outer layers. Vibrating electrons emits gravitational waves.

These gravitational waves are a lot lower energy than the radio waves you know (of the order of 10-39 smaller). But this altered gravitational field causes a resonance effect in the electrons of one atom of each pair of the BAAYIODUU (the atom which we will call, in our language: BAAIGOO EIXUUA and is untranslatable) DYNAMIC or Shifters.

In other words, it acts as a receptor capable of detecting gravitational waves emitted by cytoplasmic OXYGEN and save the message as if it were a terrestrial tape. When an electron is associated with gravity Quantum (called GRAVITON on earth) such an association may give another electron with phases and orbital position change and a new sub-particle degrades thereafter by subdivided into two.

Thus electrons KRYPTON atoms are "informed".

The oxygen atoms of the intracellular water emit gravitational frequency captured by the electrons of one of the 2 atoms of each pair of krypton of baayioduu. That's what that means decoding of baaigoo eixuua. In other words, when the electronic layer interacts with a gravitational frequency, this impacts a specific electron. This defines the flow of information from the environment to the planetary-BB.

Conversely, planetary-BB may transmit to krypton's electrons of baayioduu, subconscious information which will propagate through the oxygen atoms of the intracellular water. Information will join the neural circuit.

One hypothesis is that when the planetary-BB sends a mutation "order" the process is different. The gravitational flow of the planetary-BB captured by krypton's electrons interacts much more directly on DNA via a relay compound. This relay compound would have the property of converting gravitational frequencies directly into bio-frequencies. These bio-frequencies then realize the mutation directly at codon or DNA base with which they resonate.

THE BIO-FREQUENCIES

Can there be such a bridging compound between gravitational frequencies and bio-frequencies?

The research hypothesis is to study compounds where the total number of protons and neutrons would be equal to that of a pair of krypton atoms. We find that the molar mass of Kr_2 is identical to a compound of $GeSi_2C_3H_3$ approximately 167.6 g/mol.

This relay compound of GeSi2C3H3 would therefore do the interface between the DNA and Kr2 pairs. It may synchronize all the cellular elements required to manage the planetary-BB drived mutations. It would be the one which protect DNA against unwanted mutations and triggered controlled mutations.

Diagram of intracellular flows Krypton-DNA

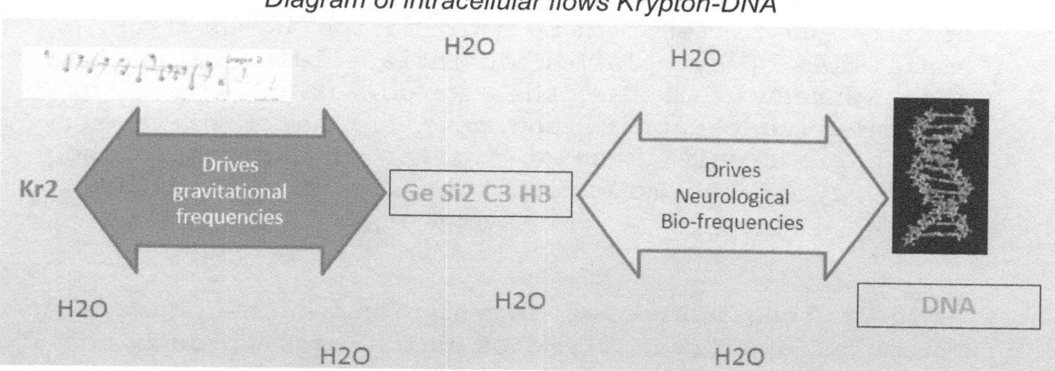

The assumption is that this compound—Germanium, Silicon in particular—thus that various crystals and the main rare gases play a key role in cellular metabolism and bio-frequencies. Many cellular processes would be under controlled by those bio-frequencies via these elements. The carbon atoms and hydrogen atoms may be the atoms interface with the bio-frequencies from neurological processes. We will develop this point in Chapter Telepathic communication.

The best known, the silicon vibrates at a very stable frequency. This feature makes silicon an excellent receiver and transmitter of electromagnetic waves. This is why it is used extensively in most electronic technologies.

D45: These are the nerve impulses that, thanks to the different carbon atoms and helium whose QUANTIC statements have been excited, modify ordinary resonance frequency state Zero (plane wave) of each KRYPTON atom by OWEEU OMWAA effect. And messages in the memory, for example, will be encoded in these atoms WAVE shaped.

D21: The pulse frequency activation of nerve centers located in the ventrolateral choroid plexus of the brain is 6,123 cycles per second (unit widely used in neurophysiology frequency).

Biogenetic constant: 658,102 x 12-10 seconds. This is the time taken by the quantic state to be established in the carbon atom of the deoxyribonucleic acid chain for the formation of a GENE.

D731: The other atom pair captures the information of the environment. This information comes from a small mass of cytoplasmic or intracellular water and also intranuclear. In other words, it is the water molecules that capture the wave trains of different lengths, not only those of similar dimensions to the frequency of the molecule, but also of metric wavelengths.

The second source of information is the chemical molecules and trace elements which pass through the cell membrane.

[...] In some universe we detected life with negentropic forms [...] with as a central element germanium and silicon.

BAA IODUHU (B.I.) or it protects from mutagenic or it itself causes a controlled mutation action.

Fish sends information on its genes and the environment, and receives only [...] genotype patterns to modulate its mutations.

According to microbiologist E. Guillé, redundant sequences of a part of the "junk DNA" would have a function as transmitters and receivers of electromagnetic frequencies which would be a possible new function for a part of the "junk DNA."

According Dotto Gianni A., in 1971, the magnetic charge of the genetic code is maintained at an appropriate level by the electrical property of the double helix, which functions as a common transformer level, where the primary voltage and the secondary winding is proportional to the number of turns of the coils. For human between 35 and 55 voltage from 45 to 70 millivolts maintains linearity of 10 base pairs per turn in the DNA double helix.

In their synthesis of the compound based on Ge Si C, J. Kouvetakis and D. Nesting from the Department of Chemistry and Biochemistry, University of Tempe, Arizona, show that Ge Si C alloys are metastable.

There are many properties related to high frequencies in particular, and sensitive to very low gamma radiation.

In 1973, Zheng Tsiang Kan seems to have transferred genetic information using a bio-electromagnetic radiation at ultra-high frequencies.

In 1991, Jacques Benveniste transfers a molecular signal using an electromagnetic sensor and of a low-frequency amplifier.

Effectors of GeSi2C3H3

Inside cells, germanium could be detected in the lysosomes, condensed chromatin and nucleoli. The physiological role of germanium is currently unknown.

Germanium is stored naturally in garlic, ginseng and ganoderma lucidum mushroom especially. Consumption of these famous mushrooms appears to be linked to the aging.

Traditionally, major Chinese and Japanese Emperors dynasties used the fungus to extend their longevity. The Chinese and Korean research also highlighted the properties of this fungus to cause apoptosis of cancer cells.

DNA telomeres play a major role in the cell aging. More telomeres shorten, cells age more.

Colloidal-type elements platinum, rhodium, palladium, iridium, etc., In their colloidal form can bind to telomeres. Geneticist Maxim Frank-Kamenetskii wrote about the DNA: “Base pairs are arranged like a crystal.”

All this, makes us think that the GeSi2C3H3 and some germanium atoms scattered in the cell, directly or indirectly influence the lysosomes, the telomeres, the nucleotide sequences of DNA and the nucleolus in the process of DNA replication.

Overall, the compound GeSi2C3H3 would ensure synchronization of all elements of the cell involved in the process of DNA replication and thus ensure the maintenance of DNA integrity, especially against the aggressions of cosmic or radioactive radiation.

Synthetic diagram of GeSi2C3H3 effectors

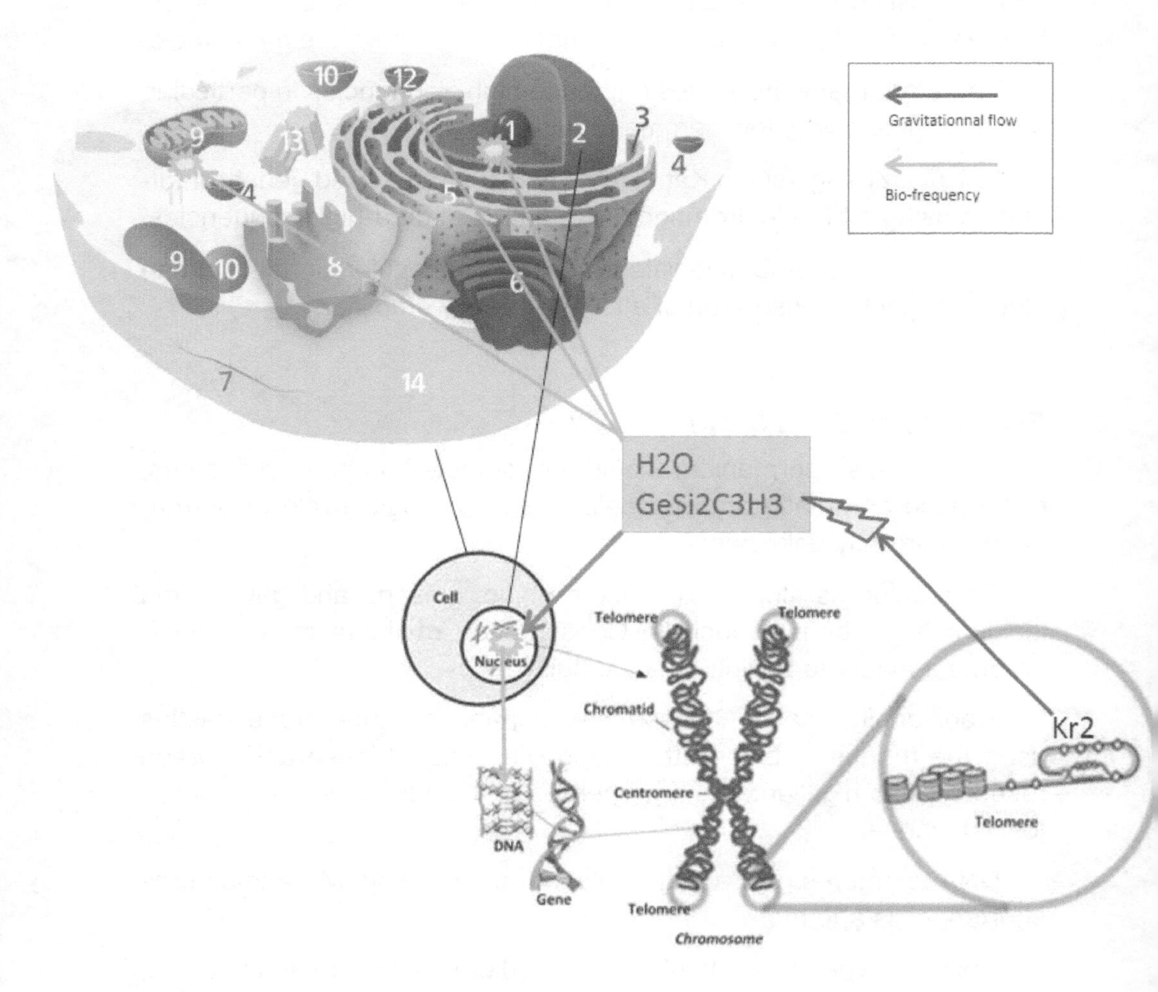

1. Nucleolus
2. Core
3. Ribosome
4. Vesicle
5 Rough endoplasmic reticulum (granular) (REG).
6. Golgi apparatus
7. Cytoskeleton
8. Smooth endoplasmic reticulum
9. Mitochondria
10. Vacuole
11. Cytosol
12. Lysosome
13. Centrosome (consisting of two centrioles)
14. Membrane plasma

Frequencies of amino acids and proteins

As we discovered through the analysis of UMMO words in *PRESENCE 2— The language and the mystery of the UMMO planet disclosed*, each of the twenty amino acids emits a wave which we can calculate the specific frequency, following the work of Joel Sternheimer, researcher at the European University Research in Paris.

These waves are emitted when these amino acids are carried by transfer-RNA to combine proteins. Similarly proteins emit a harmonic frequency resulting from amino acid frequencies. Some resonance with the sound waves can influence the proteinic cycle. This has the practical consequence that sound frequencies can affect the development of life and plants in particular.

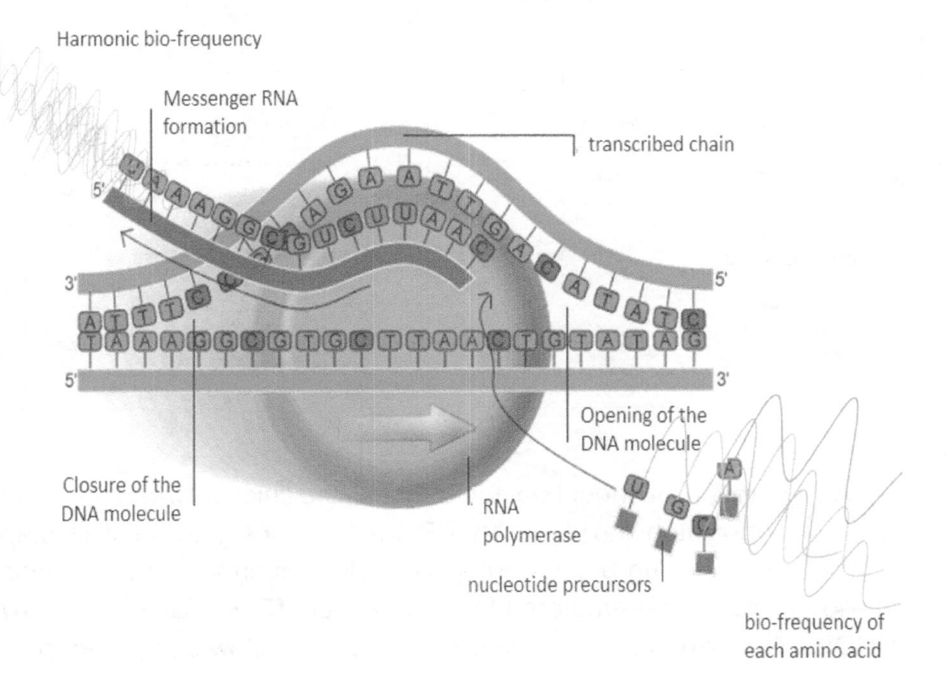

CONCLUSION

The assumption is that the planetary-BB is structured with IBOZOO tree networks in which each level of categorization is an emerging level.

Two OEMMII phyla of the same planetary-BB does not generate two separate planetary-BB, but two separate IBOZOO networks in the same planetary-BB.

Figure 1-p2: Alive beings communicate with the BB according to specific gravitational frequencies for each species that are juxtaposed to the gravitational frequencies of the planet.

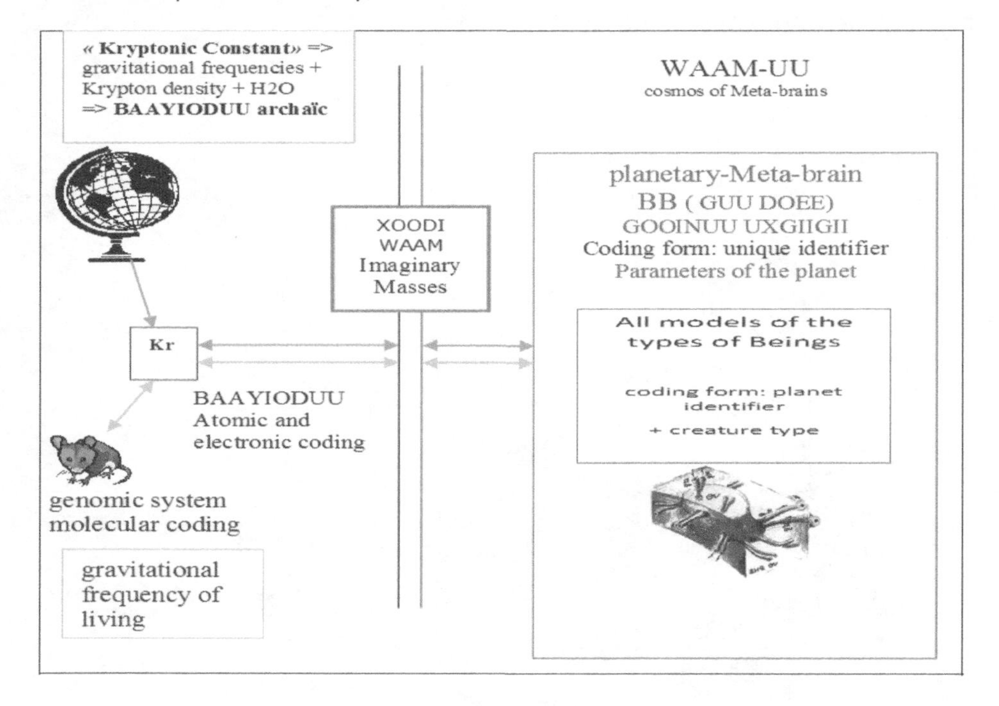

As a corollary, it should be noted that all the links between the configurations of krypton and planetary-BB are dynamics. The set of all "standard model" of living beings possible evolve within the range of what is allowed by the parameters of the planet: coefficient BAAYIODIXAA UUDIE. This dynamic evolution is obviously very slow compared to the longevity of living beings; we are here in the geological time scale.

Only planets that have rare gases, crystalline minerals and water can develop a BAAYIODUU linked to a planetary-BB. There is no life without the role of transmitters/receivers of these bio-frequency components in aqueous solution.

In the cells the water captures the external frequencies, and the Krypton-GeSi2C3H3 complex the «internal» frequencies of BB.

Cybernetic diagram of Life. The synthesis scheme of Living steering flow:

DNA <-> Bio-tuner <-> krypton BAAYIODUU <-> WAAM-UU [Meta-brain planetary BB]

Brain <-> Bio-tuner <-> krypton OEMBUAWE <-> WAAM-UU [Meta-brain planetary BB]

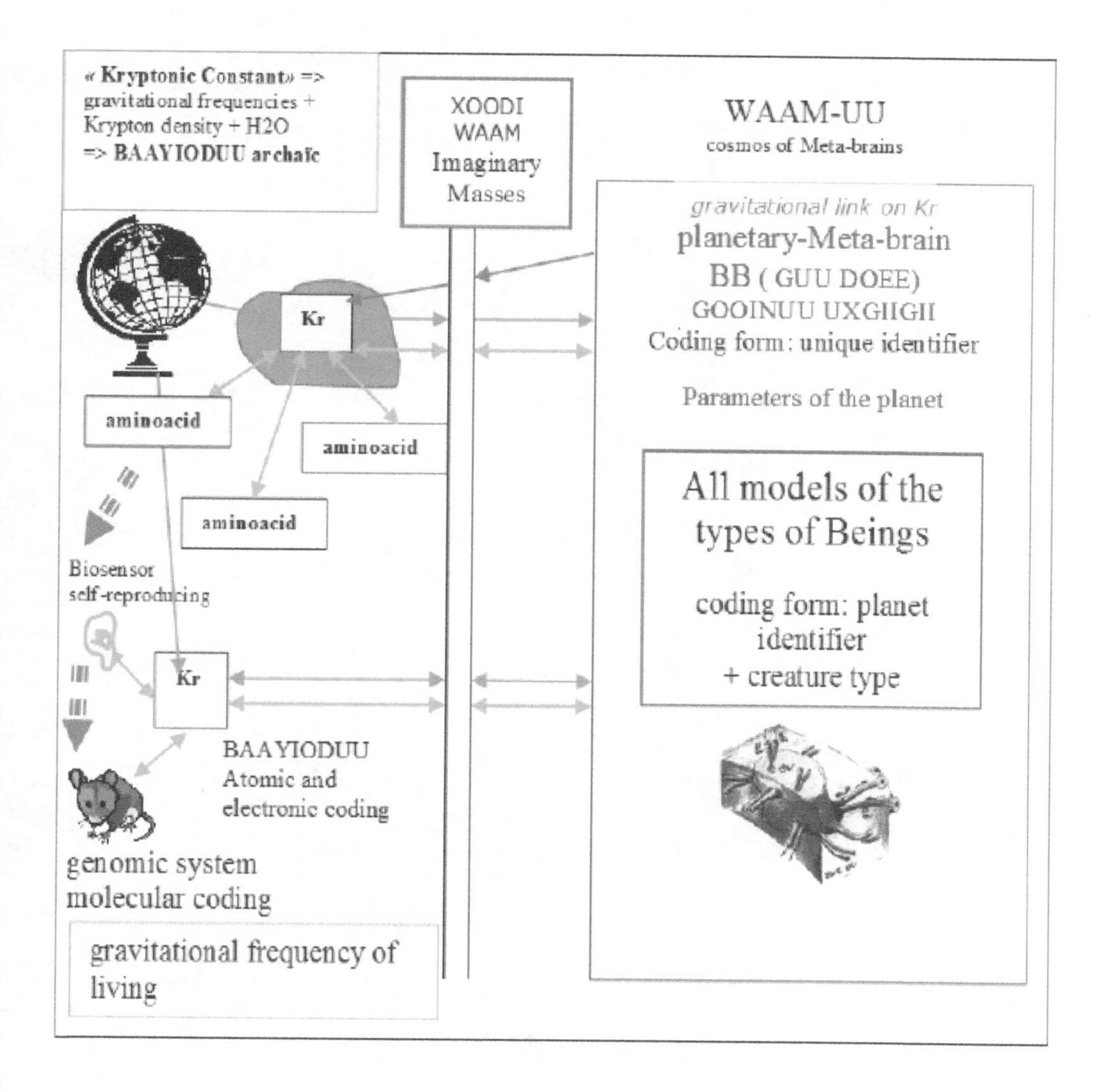

*

THE GENESIS OF "SOUL" BUAWA

We saw in the previous hypotheses how, given a universal constant underlying so-called "Kryptonic constant" life could emerge from the inert matter from the formation of an archaic-BAAYIODUU. Then, when integrated with alive beings, how can be established communication flows between entities. We now consider the connection to the Soul BUAWA.

HYPOTHESIS ON THE FLOW OF INFORMATION

The WAAM-U consists of "infinity" of BUAWA. The BUAWA are initially "empty".

The assumption is that the content of the BUAWA is generated precisely when the chromosomal fusion of fertilization; simultaneously the BAAYIODUU connection is setting with the BB.

In this chromosomal fusion, each haploid gamete, male and female, carries a series of 86 krypton atoms from meiosis. Karyogamy, assembly gamete nuclei also cause the assembly of pairs of chromosomes and therefore pairs of krypton atoms. A new BAAYIODUU is created in the first diploid cell of OEMMII, thereby causing the effect of the LEIYO connection commissioning with BB.

This has the direct result that human twins might initialize and use a single BUAWA.

It is in the human embryogenesis and a priori OEMMII in general, during the structuring of the brain in particular, that OEMBUAWA is taking place. The establishment of OEMBUAWA activates the communication with the BUAWA and with the BUAWEE BIAAEI (the collective soul of OEMMII). For beings "lower" which have no OEMBUAWA, so there is never BUAWA activation, but activation is still potential, if the living being evolves.

The principle of identifying the OEMMII described for BAAYIODUU link (BB-krypton-living being-genomic system) is identical to the link OEMBUAWA to BUAWE BIAAEI the collective soul of OEMMII. This link includes de facto the planetary identification. It is universal, multi-cosmos, specific to each OEMMII and permanent until the death of the individual. Note again that in this case, the twins necessarily have the same BUAWA enriched with the information of the two individuals.

The identification of the OEMMII with its BUAWA, is due to the transmission of an energy quantum consecutively to a precise gravitational frequency as the OEMBUAWA-BUUAWE BIAAEI link. The gravitational resonance between the OEMBUAWA and BUAWA corresponds to a "mere" alignment between the axes UU IBOZOO carrier of gravitational information and the chain of the IBOZOO UU of the BUAWA. The propagation of gravitational wave by "subquantic" resonance effect, is a LEIYO effect transmitted through imaginary masses who do not generate mass, yet can record information in the IBOZOO UU chain of BUAWA. Conversely, the information will be transmitted to the BUAWA towards OEMBUAWA following the same principle.

The BUAWA is composed of a network of "pure" IBOZOO-UU that is "formed by large chains of angular relationships. These large chains in turn form an extended substrate or matrix where is recorded all the information of our lives in a part of this network" and another area of the network of IBOZOO UU that "codifies a whole program statements, that conforms each OEMII (man took in his neural dimension)." The OEMBUAWA-BUAWA link also corresponds to a specific gravity resonance equal to that initialized the BUAWA content when genomics merged and set so, in biunivocal and dynamic way.

It seems that the "free will" is a human decision process involving several potential solutions. It is performed at the brain. The BUAWA do not make decisions, it generates a "guiding principle" that conforms to the "psychic profile" constituted by the IBOZOO network of BUAWA and could undergo various external influences (see "Assumptions about the influence of planetary configurations on the psyche ").

Then this "guiding principle" is transmitted to the brain that confronts it to perceptions of the environment, and confronts it to mental models transmitted by BB. As shown by our oommo visitors, the "free will" is simply the decision among all possible choices with approximately 70% of the decisions in accordance with the "guiding principle" of buawa. These consciousness-subconscious processes are expressed by the words of eese family by oommomen *(see oommo dictionary).*

Soul and Time

The "guiding principles" of the soul-BUAWA are stored data. The flow of this data from outer space is through various cosmos structurally different.

In the planetary Meta-Brain speed photon is infinite. In the Soul-BUAWA, there is only one dimensional axis that stores information. The concept of photon speed or time has no meaning in this cosmos. It's simply storing information in chains of IBODZOO like computer bits of terrestrial computers of our time.

On Earth, in the course of our three-dimensional universe, time is also composed of a chain of discrete elements of IBODZOO. The stream that connects the soul-BUAWA to the human body via the connector OEMBUAWE krypton's atoms, joins the discrete IBODZOO elements of the 2 channels.

The flow from BUAWA, passes into the cosmos XOODII inter-layer photonic whose speed is infinite.

Then the infinite flow velocity reaches the OEMBUAWE antenna in our three-dimensional framework, governed by our time that consists of a chain of discrete elements of IBODZOO where the Planck time is about 5,391 x 10^{-44} seconds, corresponding to an elementary angle on the time axis between two "multidimensional nodes" IBOSDSOO.

That is to say we'll move from an infinite velocity flow, to a flow at the speed of light, our 300,000 km/s.

The communication between the soul-BUAWA our brain is almost instantaneous. This is also the case with telepathic interpersonal communications via planetary Meta-Brain BB.

D731: The information of our mind is also transferred to the psyche. There, it is recorded on filamentous networks IBOZSOO UHU. That is to say on IBOZSOO UHU chains. Likewise this "wireframe" appears on the UI sequence that interacts with us, directs us. Each of these chains I.U. is composed of infinity (in the physical sense) of angles which encode information.

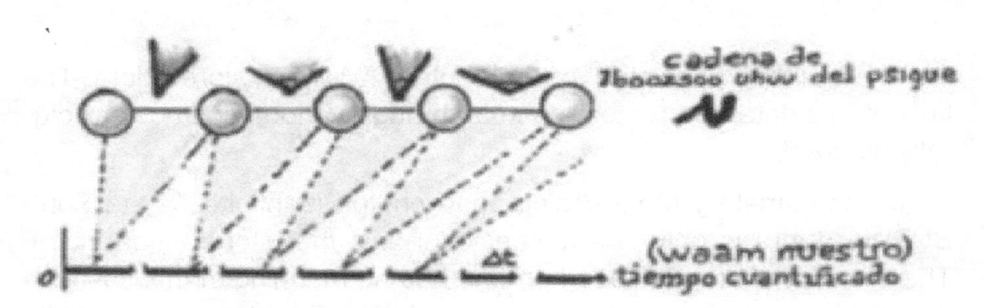

There are, as you can see on the chart, one correspondence between the moments of the time axis [the time is unified as discrete (NdT: composed of separate elements)] and IBOZSOO UHUU of the psyche. The time in the universe is formed by a succession of discrete TEMPORAL quantum Dt, each of which is related to the pair of IU that encodes the instructions that soul sends.

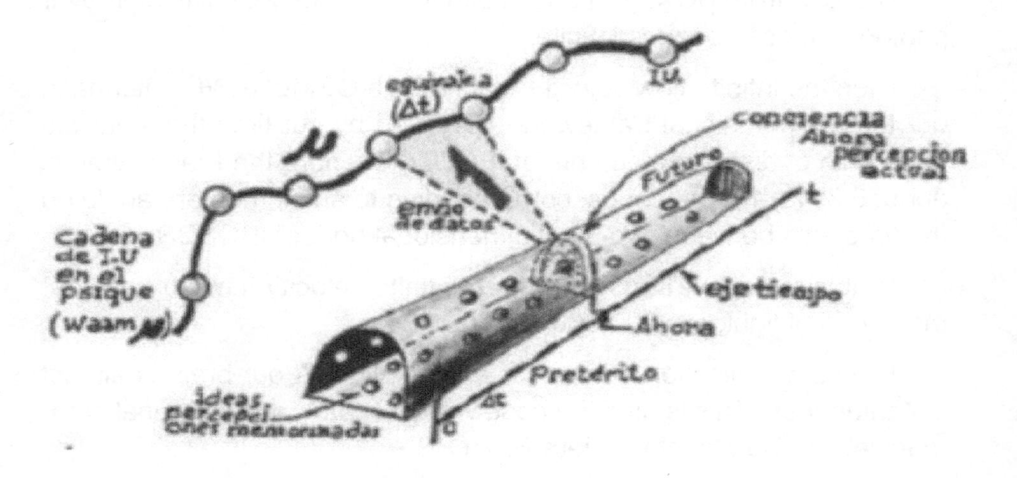

PROVE THE EXISTENCE OF THE SOUL BUAWA

It turned out that in the case of the decision of a motor action, the decision of the action is taken in the frontal cortex and then sent to the parietal cortex, which generates a "potential readiness" BEFORE conscious decision-making.

In this case, an unconscious process of the frontal cortex is identified as the maker of the motor action. The conscious decision is simply being reduced to a "GO/NO GO" in the motor cortex, after the initial non-conscious decision.

The first experiments showing, incidentally, that unconscious processes of our brain decide before us, were carried out in 1983 by Benjamin Libet and confirmed in 2003–2004 by Angela Sirigu and Patrick Haggard.

Where would come the subconscious decision? The hypothesis of BUAWA explains the origin of the frontal cortex subconscious decision...

Diagram of OEMBUAWA-BUAWA information flow, the process of a motor action decision would be:

BUAWA—> OEMBUAW—> frontal cortex—> parietal cortex—> motor cortex—> consciousness—> motor cortex—> Action

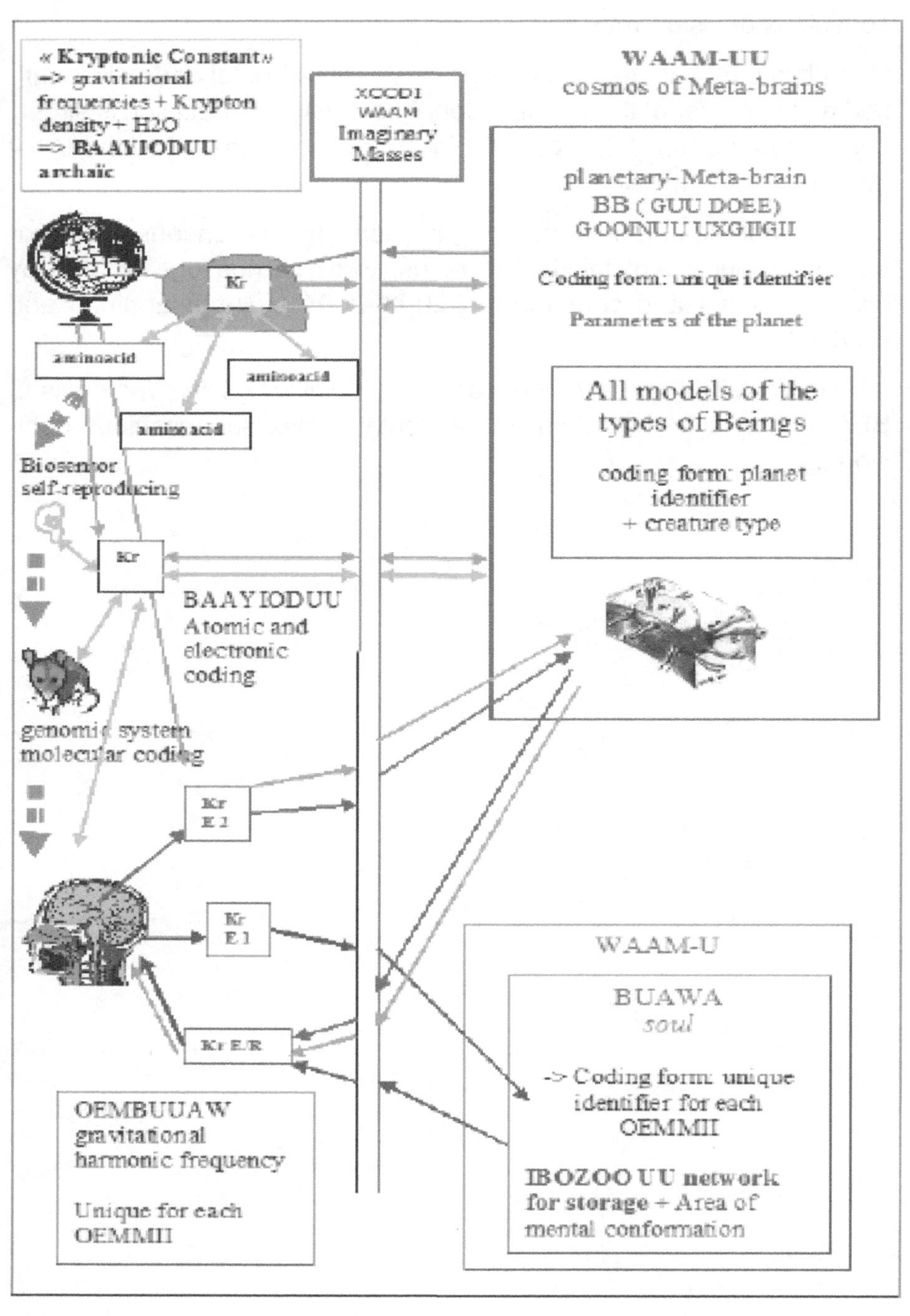

THE CONCEPT AND THE LIMITS OF "REINCARNATION"

Usually the metaphysical concept of reincarnation is the integration of the soul linked with a human body to the body of an animal. The simple being disparity leaves us easily assume that this is neither consistent nor realistic. But, what about reincarnation in another human body?

I will begin with a personal testimony. A friend of mine, the late Gérard P. who contributed to the scientific adventure of the Presence series, had a particular psychological profile, he was diagnosed with Asperger's autistic. Curiously, in his younger years, the child Gérard showed atypical behaviors and skills, one of them was to know "innate" mathematical equations. Years later, he sought to what were these equations and found that the equations were formulated by Alan Turing, who died the day of his birth... The coincidence then appears as a mental transfer between the dying person and the newborn. As we shall see in detail, one possible interpretation is that some of the information generated by Alan Turing into our planetary Meta-Brain BUAWE BIAEI, were transferred to it in the soul-BUAWA of the newborn Gérard P.

Numerous observations and scientific studies have been done on this topic. These include studies James Parejko, professor of philosophy at the Chicago State University, on hypnotized subjects, the work of William E. Cox, Joel Whitton, professor of psychiatry at the Toronto Medical School and Joe Fisher which several subjects manifested knowledge of ancient and rare languages without even assumed the prior existence.

Still other people identified in detail and proven places, or even people, from their "previous life". As for Ian Stevenson, professor of psychiatry at the School of Medicine of the University of Virginia, he studied numerous cases where the subjects had morphological manifestations, bodily marks related to the violent death of the previous incarnation...

Now examine possible cases related to reincarnation. In my understanding of Oommo documents, should be considered in the design of a human being by chromosomal fusion, two special things can happen:

• Exceptionally, the identifying information of the soul of the new human corresponds to an existing BUAWA soul. And so, the new individual will have a connection to an informative heritage of the previous individual.

Although there are endogenous psychic phenomena to the brain, more or less psychopathic and phantasms, the feeling of having "experienced something in a past life" may not be so unfounded... Some individuals might well have inherited very old and very rich in accumulated information souls, back to the earliest Homo habilis.

As a corollary, we can ask ourselves if, in addition to the cultural context, individuals having a BUAWA with richest information content are also humans who have the most developed capabilities prospective?

• A new human connects to a soul-BUAWA empty cell. This is quite possibly the general case. But in some special cases on the initiative of Meta Brain BUAWE BIAEI, it decides to transmit specific information content to the empty BUAWA. We can imagine that this action aims to modify the human evolution directly via the social network.

We perceive through this view that human free will is even more limited than we often realize. The human seems to act freely, but by drawing his thoughts with information contents already filled, the work of neuro-brain is making arbitration and produces a thought adapted to the physical environment of the human...

H3– The new soul is born can be impregnated ... it may receive information proceeding from the collective Soul. In this sense the information it receives proceeds of other souls. In that sense, yes reincarnation would be eligible. This somehow, it is information from Collective Soul.

H4– the resurrection can happen, you know your universal consciousness persists in another brain. This is not a reincarnation. Consciousness can persist into another body (when the characteristics of the individual coincide, like a glove can become another hand).

We see how the human body point seems to be little, in comparison to the influence of cosmological objects "metaphysical" that guide its actions. As we have seen, is the meaning of the Oommo word BUAWA: Generator Interconnection actions.

An additional point would be to consider the human being as a cumulation of his body dimension and his extra-corporeal dimension, that is to say, with his soul-BUAWA.

This is what we discussed in "Presence 2 The language and the mystery of the planet UMMO revealed" with the analysis of the meaning of OEMMII indicating the association of OEMII physical body with a "border" that would be the soul—BUAWA. This can be summarized by:

OEMII (physical body) + BUAWA (Soul) = OEMMII (human)

BB's IMPREGNATION ERRORS

The dissociative identity disorder or disorder of personalities is known as a mental disorder defined in 1994 following a set of diagnostic criteria as a particular type of dissociative disorder.

It involves at least two individuals who systematically take control of the individual behavior with memory loss beyond the usual oblivion. Some patients may have 10 separate personalities, identifiable by distinct mental maps, specific to each personality disorders ranging from allergies to diabetes... Each of these personalities is in itself normal...

It is as if one physical body had been designed with several Souls...

The assumption is that the impregnation of the BB's data in the Soul to chromosomal fusion time, went wrong. BB would have impregnated the soul with multiple data streams from different people. Probably by segmenting different areas BUAWA instead of mixing all the data in the same area... So, the soul connector OEMBUAWE, the brain of the person, alternately accede to these different areas...

INTEGRATION OF THE SOUL TO THE PLANETARY META-BRAIN

On the death of a human OEMMII, occurs an agglutination phenomenon of cosmic entity BUAWA with its planetary Meta Brain in the WAAM-UU cosmos.

This match between these two cosmos entities so different requires the implementation of an inter-cosmos relationship. The link between BUAWA one-dimensional and planetary BB penta-dimensional is done via a IBODSOO chain. The deceased's Soul-BUAWA can then exchange information flows transiting the planetary-BB. This allows communication with all other Souls-BUAWA of deceased, and potentially with OEMBUAWA effectors of human brains.

It is also possible that this integrated Soul-BUAWA access to other BB-planetary and all other Soul-BUAWA integrated therein. After the physical death of humans OEMMII of the cosmos, all their soul-BUAWA can communicate with one another through the BB.

The integration process of soul-BUAWA is more or less long and integration more or less "complete". I express here the hypothesis whether directly under the control of the cosmogonic WOA generator. It was it who would decide the relevance of the degree of integration of the flow of Soul-BUAWA in a BB-planetary.

We believe that this process of integration of the Soul-BUAWA can also undergo more or less strong disturbances at death, because of phenomena called "paranormal" as we will explain in the following chapters.

D731: DEATH (ESCHATOLOGY OF UMMO) When occurs destruction of the last krypton elements networks (not the annihilation of the atoms but the links or network nodes), death occurs. This annihilation coincides precisely with the disintegration of certain neural networks of the brain. (cardiac arrest implies the absence of blood supply, lack of supply of oxygen and glucose to the neuronal network histology, tissue degeneration and death).

The death of the OEMII coincides with the disintegration of the OEMBUUAAW (the krypton atoms return to their quantum behavior) the BORDER EFFECT GOES therefore, appears and a fourth EFFECT "leeiyo WAAM."

A I.U network fits between two adjacent WAAM: WAAM-U and WAAM-UU.

B. B. and Soul connect to each other. This means, as we will reveal in another report that our psyche reaches the maximum integration stage in the collective psyche.

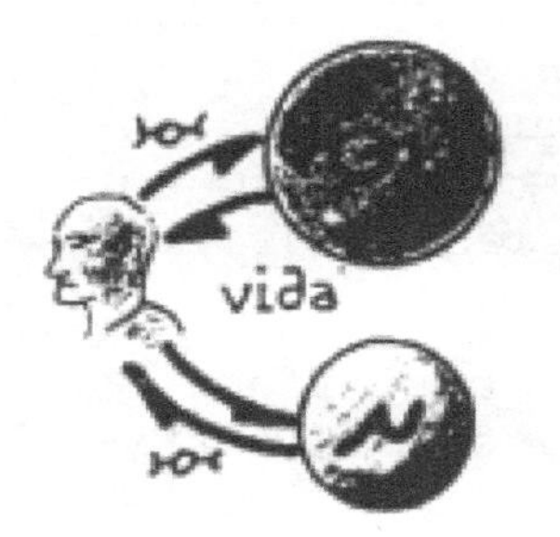

This is the sense of transcendence on UMMO. We know that when our death will occur a merging, an integration, a close liaison of the psyche, of our "spirit" (nor hardware, nor intangible but all the information matrix of our life) with the "universal" collective psyche.

We can connect us more with loved ones, communicate with the spirits of the other deceased brothers, participate in the global knowledge of all the biosphere, not only OEMII who just died, with all humans but even since born life on OYAAUMMO (and, of course, for you, from Homo habilis to the last of your brothers).

Is also possible knowledge of the real world including living beings, since BB is informed of the whole process of living beings that are not yet dead.

This means that the deceased OEMMII through his psyche can somehow influence his most loved ones through the unconscious and, to some degree also the things around them, since the biosphere changes the surrounding physical environment through living beings.

B. B. is the collective Mind. We can also call it subconscious or unconscious collective, since its contents are operational but are not made aware of us other ALIVE BEINGS.

The mind of a deceased brother can be, and in fact sometimes does, we attend, we protect and sometimes interacting so VERY ACTIVE, but most of the time, gently modulating our unconscious through information we receive from BB.

The psyche or soul, freed from the tie entity => BUAWA and OEMII (or physical body)—already disintegrated—begins a joyous step eternal knowledge of BB, not only to assimilate a Millennium culture accumulated by centuries of life of all human beings, but it will enter into the knowledge of science, art, in short the whole culture of planetary humanity.

It probably also will feel the pain, but offset by the deep knowledge of the UUA, moral lives and euthymic beings.

Moreover: as a participant in the WAAM, it will have access to the eternal secrets of all the WAAM-WAAM, assistant to the constant evolution of its galaxies, stars and various mass formations.

D357: At the time of death, O., that is to say, the krypton atoms, cease to hold office. But instead, B. (soul) is completely connected through valves between the two WAAM (WAAM-U and WAAM-UU) so that it amounts to a genuine integration of the soul to the collective soul, where she participates in all the knowledge accumulated by humanity.

This is our scientific knowledge of transcendence after death of a OEMII.

A network of IBOZSOO UHUU acts as a valve between B. (SOUL) in the WAAM-U and BB inserted into the WAAM-UU, enabling "integration" almost absolute between the two entities. It's WOA (God or GENERATOR) that defines the characteristics of this chain UI (Information valve) into a "time" determined.

If OEMMII in areas where he is responsible and free, throughout his life violated the laws UUAA (ETHICAL), it is necessary to transform the coded information of its structure in BUAWA. Remember that the SOUL do not think that it is a simple frozen data matrix. She can handle its own heap of information only with the help of BB.

The BUAWA psyche may be doomed to suffer from slow "capacity" to use his own EGO (information encoded within it) and does not participate in the dense complexity of BB.

But WOA may, if the man has met the moral standards during its existence or after the correction of his structure once he died (reconforming), allows this UI Network to offer him an exceedingly denser flow of communication that we experience in the course of our existence as living beings in our WAAM.

In this case, the "integration" of BUAWA (SOUL) in the BB is so intense that it shares the immense volume of data COLLECTIVE SOUL. His intellectual vision of WOA (God) increases. It penetrates deep knowledge of the Cosmos, of the evolution of creatures, the vast "knowledge" (intellectual and emotional information) contained in the BB.

Observe that, in some way, this notion coincides eschatological, with some accuracy, with the theological estimate of Christianity OYAAGAA on Salvation.

What you call Purgatory is in this case the Reconforming process, which reduces to the fact that WOA limits to some degree of participation of B. in BB, reducing to different degrees the value of the Canal or valve that separates the Both WAAM (WAAM-U and WAAM-UU).

What you call "GLORY or Salvation" is the full integration of the Soul, not exactly in God, but in a grandiose creation of WOA as is the BB (COLLECTIVE SPIRIT). We can imagine the wonderful "ecstasy" or "enjoyment" that our mind can experience not only by the fact that information "registered" in him to be treated fluidly (the mind by itself could do), but also by participating in and benefiting from the vast ANY information contained in the WAAM-WAAM.

Through the BB he can communicate with other BUAWA of his deceased brothers, and like every BB participates in the printed information matrix in the WAAM-UU from the moment of its creation or generation (Remember WAAM-UU aims to meet the peculiarities of the whole WAAM-WAAM.), his spirit will enter into the most intimate secrets of the Multiplan Cosmos (the Universe).

EXCERPTS FROM GR1-4, CORRECTED TRANSLATION, TEXT REWORDED AND COMMENTED, FROM THE ORIGINAL.

Note du destinataire: "Les traits et les couleurs du dessin ont été changés et peuvent ainsi permettre d'identifier à l'avenir, s'il n'y a pas eu de fuite de l'original, celui qui se revendiquerait être l'auteur de ce dessin"

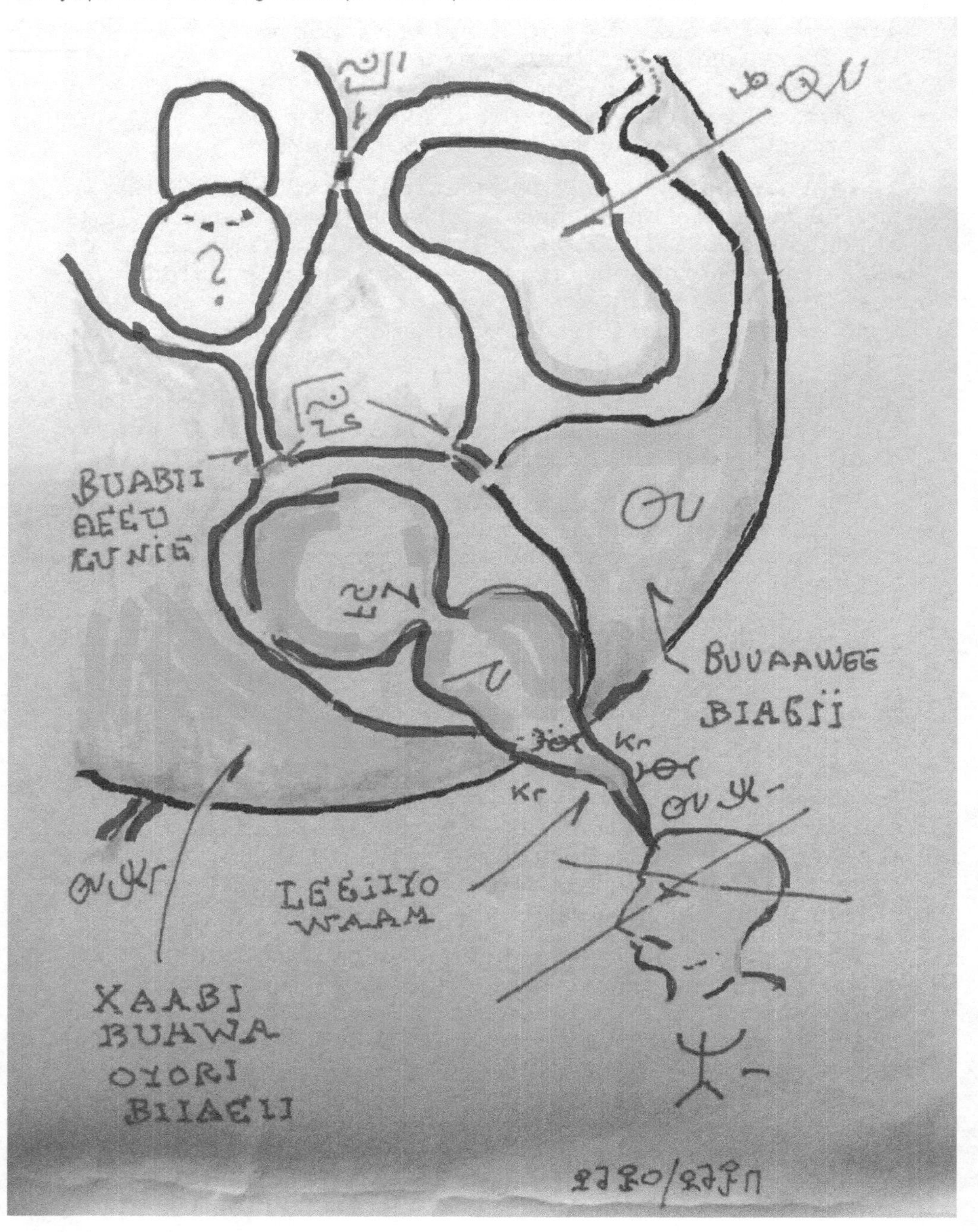

When the OEMMII dies, BUAAWA (actually a small number of Krypton atoms located in the columns of what you call the Neocortex) «launches» a signal to BUUAAWAA BIIAEII, which, thanks to (a boundary effect) LEEIOO WAAM, processes these atoms by transforming this set of elements of physical matter with electric charge and information carrier, into pure units of information (the boundary effect allows the electro-chemical coding to pass into «bits» of information into another cosmos)

The transmitted energy is incorporated into the filaments of BUAWAA BIAEII as an energy addition.

...

The BUAWWAA of the OEMMII remains in the state of what You would call «permanent latency» and this will be until the death of the OEMMII who carries it, in the BUUAWAA universe (GR1-4-1), charged with PRIMITIVE information (this is very important) - primitive meaning an initial specific transconnection, of IBOZOOUU which takes place only once - at the time of conception - in the life of the OEMMII. This initial configuration of IBOZOOUU initializes the contents of the BUAWA.

We call this special nature of information: BUAWAA AMIEAYOO WADOXII: There is a «state of information», in the form of a «package» of everything that the OEMMII will do or feel in his life with his sensations, but also with his thoughts, ideas, intentions, desires, etc., although the OEMMII has recently been created, and this information is transmitted in one go to WAAM BUAAWAA (GR1-4-2).

What is for You is a great mystery, yet very real, and was expressed mathematically on UMMO long ago. We must confess, however, that this has been experimented with statistically with probabilistic mathematics and with enormous proofs of success, but not completely irrefutable. We at UMMO take it, however, as a provisional scientific paradigm.

...

When the OEMMII dies, the information packet that constituted OEMMII (a priori augmented by the information transmitted throughout its life), is «projected» into BUUAAWAA BIIAEII via the LEEIOO WAAM boundary effect.

BUAAWA BIAEII then retrieves the information contained in the BUAWAA of what this OEMMII was, AND COMPARES IT WITH THE ARCHETYPE (WOAIYIIBUAA) THAT BB HAD PREVIOUSLY *PREPARED* FOR THIS PARTICULAR SOUL

establishing from this moment a permanent link between the said BUAWAA and a cell or «niche» (XAABII BUAWAA OYORII BIAEII) which is found in BUUAAWAA BIIAEII and which contains a kind of « mirror BUAWAA « as we called above (which is only a specific ideal BUAWAA created directly by WOA in compliance with AIIODII)

to which the homogeneous groups of information contained in BUAWAA will pass in discrete quantities (e.g. a speech delivered in one's life by the OEMMII, or a hand-to-hand struggle with a brother, or a copulation between GEE and YIEE).

The said «niche» has a «device» or BUABIIAEE ULUNIE that serves to transmit the information contained in other niches comparable to the cell in question - it is BUAAWEE BIAEEII that decides the flow and timing of this information.

When the information contained in the archetype intended for that soul becomes equal to the information in that cell, we say that the soul is reconformed.

In a simplified way, we could say that at the chromosomal fusion of a sperm and a human egg, the BB initializes a BUAWA Soul with a packet of information. We could call this packet of information the DESTINY, or ROADMAP, of our entire planned life.
Nevertheless, the DESTINY, or ROADMAP, can be modified with events due to the chance of physical matter that escapes the control of the BB and in part by free will. The individual increments and completes the information of the BUAWA Soul throughout his life.
In addition, BB created for him a duplicate of the initial BUAWA, the mirror BUAWA.
And in his structures, BB has models, archetypes, BUAWA WOAIYII-BUAA that he will need to evolve the humanity he manages.
Upon the death of the individual, BB connects the BUAWA to a

«niche» that contains the mirror BUAWA. BB manages multiple transfer devices and controls the information of these types of objects. BB filters the incoming information into the «niche» and de facto into the mirror BUAWA, aligning with the intended BUAWA WOAIYII-BUAA archetypes .
When the «niche» and de facto into the BUAWA Mirror Augmented, is equal to the WOAIYIIBUAA archetype of the planned BUAWA, then the BUAWA Mirror Augmented is operational, the UMMO's people say that the individual's BUAWA has been reconformed (in the BUAWA Mirror Augmented).

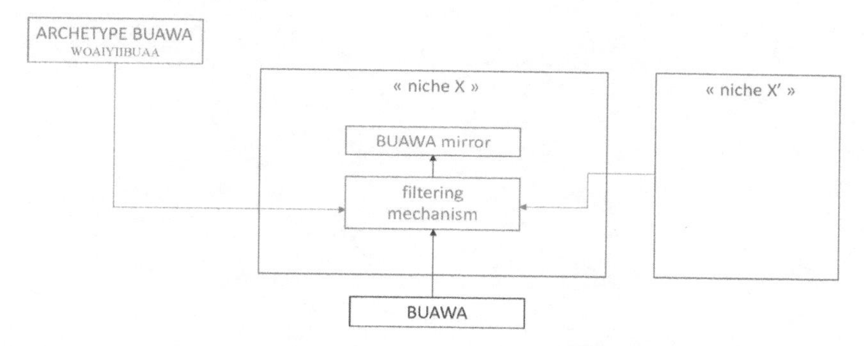

Upon the death of the individual, BB connects the BUAWA to a "niche" that contains the mirror BUAWA. BB manages multiple transfer devices and controls the information of these types of objects. BB filters the incoming information into the "niche" and de facto into the *mirror BUAWA, aligning with the* intended BUAWA WOAIYIIBUAA archetypes .
When the "niche" and de facto into the BUAWA Mirror Augmented, is equal to the WOAIYIIBUAA *archetype of* the *planned* BUAWA, then the *BUAWA Mirror Augmented is operational, the UMMO's people say that the* individual's BUAWA *has been reconformed (in the BUAWA Mirror Augmented).*

*

THE INFLUENCE OF CELESTIAL BODIES ON THE MIND

We have seen in previous hypotheses how the planetary-BB and buawa were initialized, and what are the flows that could be involved. So the buawa generates a "guiding principle" which conforms the "psychic profile" and consists of the ibozoo network of buawa and could be subject to various external influences.

Paradoxically, on this issue we have different verifiable elements and yet it may seem like more speculative and more disruptive than previous issues. This is due to the fact that in addition to experimental difficulties, the supposed politically correct search "ill-treated" in both senses of the word this issue. Researchers so far have dared approach it can be counted on the fingers of one hand. So i had few friends in dealing with this matter likely to do much ink...

Point out, if necessary, the newspaper horoscopes have no resemblance to real planetary configurations, and contribute to maintaining the taboo. The origin of the taboo is that the ancient empirical systems described the relationship between the psyche and Ptolemaic geocentric vision of the planets. By discovering the heliocentric nature of the solar system, most of researchers conclude too quickly that the ancient empirical systems such as the zodiac, were also wrong. In fact, it is obvious that one can easily calculate the geocentric view of the planets from the actual configuration of the heliocentric planets but "the baby was thrown out with the bathwater."

State of the Art

About gravitation, to date no graviton was found. The only thing we know is checking the manifestation of a force linked to the masses.

Facing this ocean of unknown nature of gravity itself is very mysterious, and all that can be said about the gravitational "frequency" is very speculative. I take it that these waves have at least the common properties with known waves.

Among the few serious works on the subject, one can note two distinct hypotheses. The first hypothesis developed in 1974 by D. Verney assumes these are the gravitational effects of planetary configurations that resonate with the psyche and the second presented in 1988 by P. Seymour, supposes that electromagnetic phenomena are produced by the planetary configurations affecting the psyche.

In 1990, following the work of Verney D. and R. Penrose, I extended these approaches imagining that there must be "a gravitational resonance effect of the stars on a quantum brain effector." This speculative thought was expressed in 1995 without knowing the Oommo documents and therefore without being able to develop this idea...

In this chapter, therefore we will develop this idea with lighting of the Oommo documents.

The context of the assumption

Regardless phenotypic characteristics of brain structure influencing behavior, we will examine the hypothesis of additional factors that could influence the behavior and decisions made by a human.

> D57-3 | T1B—13/19: "We did not know the value of coefficient BAAYIODIXAA UUDIII (untranslatable: terrestrial biological science has not yet developed this concept so important). This is a formula that expresses the biological equilibrium conditions that are measured in a given environment. Each OOYAA (Planet) has special conditions that allow or not the existence of a carbon biological cycle in the troposphere. The biogenetic development of the morphology of animals and plants will be a function of a series of physical constants. «

"The formula the BAAYIODIXAA UUDIII expresses is a complex function that involves using a multitude of parameters such as: Acceleration of Gravity, ozonation of the atmosphere, Gamma radiation intensity, pressure and atmospheric composition, solar radiation spectrum, gravitational cycle of neighboring planets and satellites, atmospheric electrostatic gradients, telluric electrical currents, etc, etc ... which with the composition (percentage) of chemical elements of the planet, predicts the evolutionary direction of living beings regardless of other factors that may affect this evolution such as radiation which cause mutations and self-selection by the unpredictable influence of the environment."

Electromagnetic hypothesis

Electromagnetic influences on the human body are the more familiar. Their impacts relate to biological processes, biochemical, neurological, at the molecular level but not at the quantum level.

The influence of electromagnetic phenomena on the psyche, is possible on the neural structures, but outside the mental processes themselves.

Furthermore, electromagnetic effects that alter the biochemical processes of the body can be pathological, a priori it is not the case for the impact of micro-gravitational phenomena.

As for the impacts on electromagnetic sensors variations of magnetic fields, such as magnetite crystals in the cerebellum, for example, they induce punctual behavioral changes, as any other type of perception.

Electromagnetic phenomena impact biological processes, but we have not identified structuring impacts on the psyche for considered energy.

Gravitational hypothesis

- We know from the Oommo documents, that OEMBUAWA is a gravitational effector. It emits and captures various gravitational flows, which I presented diagrams in previous documents. We have also seen above, this is at the time of the chromosomal fusion that BAAYIODUU connects to BB, and then living beings communicate with the BB by specific gravitational frequencies. At this time also, that takes place the generation of Soul-BUAWA's content. Inside it, a shaping zone is created by the gravitational flow in a unique way, peculiar to each individual. It conforms OEMII's behavior, that is to say it contains the "model" of the human psychic profile.

• Moreover, we know that life evolution coefficient, the BAAYIODIXAA UUDIE can be calculated using a formula with multiple parameters specific to the planet. The "gravitational cycle of neighboring planets and satellites" is a parameter involved in the evolution of life.

• The hypothesis is therefore that the structuring of the psychological profile, that is to say, the initialization of the "psychic conformation zone" in the BUAWA which is carried out once and for all and in its completeness, and may be influenced by three factors:

• the identifying information of the planet

• the establishment of the genomic system

• other parameters that can change the gravitational flow when generating the psychic profile, at the time of chromosomal fusion

THE FIRST STRUCTURATION FACTOR

The genomic system "classic" is the main factor in structuring the psychological profile. It's linked to the archetype profiles is the Meta-Brain BB and it models the psychological profile into Soul-BUAWA by a specific gravitational frequency at the genomic fusion.

SECOND STRUCTURATION FACTOR

The structuring phase

Among the structural parameters related to BAAYIODIXAA UUDIE, there is one that is of gravitational type. This is the "gravitational cycle of neighboring planets and satellites". My hypothesis is this parameter that contributes, in part, to "shape" the "psychic conformation zone" when genomic fusion happens.

Exogenous gravitational sources from the current planet, so the neighboring planets in particular, juxtapose their gravitational waves to gravitational pulses, which are produced at the time of the genomic fusion.

All those gravitational waves corresponding to the initial impulses constitute a gravitational "harmonic" therefore in phase with the base frequency, but with different information.

All of the information is thus transmitted and generates psychological profile in the BUAWA's "psychic conformation zone", once and for all and its completeness.

So, it is the planetary configurations that contribute to "shaping" the "psychic conformation zone" of each OEMMII. The information value of the astronomical configurations is resulting in values of their gravitational forces.

More specifically, it is the differential of these values with respect to the gravitational field of the OEMMII planet which have information.

The dynamic phase

Once the psychological profile is "modeled" in the BUAWA so in part by the gravitational influence of the planets, embryogenesis continues. When the BUAWA link is activated at the establishment of OEMBUAWA, the "psychic conformation zone", then the BUAWA sends messages to OEMBUAWA "modeled" according to the gravitational harmonic which initialized it.

As the OEMBUAWA is a gravitational effector, it captures everything, like a radio antenna. So it also captures the gravitational differential of planetary configurations in relation to the gravitational field of the OEMMII planet.

Messages reaching the OEMBUAWA will be "in phase" with the gravitational differential created by the astronomical configuration of the moment, or in "phase opposition". In fact, there will be an infinite number of possible intermediate values, probably with resonant thresholds.

The messages "in phase" will have a biggest "weight" onto the brain, into the decision center, which will arbitrate more in their favor.

This will result in behavior highly consistent with the psychological profile of OEMMII.

Conversely, the messages in "opposite phase" will have a lower "weight" in the decision center of the brain, which will arbitrate more against them. The result is a tendency to inhibition of non-compliant behavior in the psychological profile of the OEMMII. These processes will be relayed at the molecular level by the endocrine system.

Conclusion

The hypothesis is that BUAWA contains a Human psychic conformation profile that the planetary configurations may interfere with.

Planetary configurations may interfere by gravitation effect on Man-BUAWA information transmission channel.

If we retain the assumption that identical twins have the same BUAWA, so in the behavioral study of twins that we can find additional information, going on this way...

So with regard to the hypothesis of the influence of planetary configurations on the psyche, the synthetic scheme is presented below against:

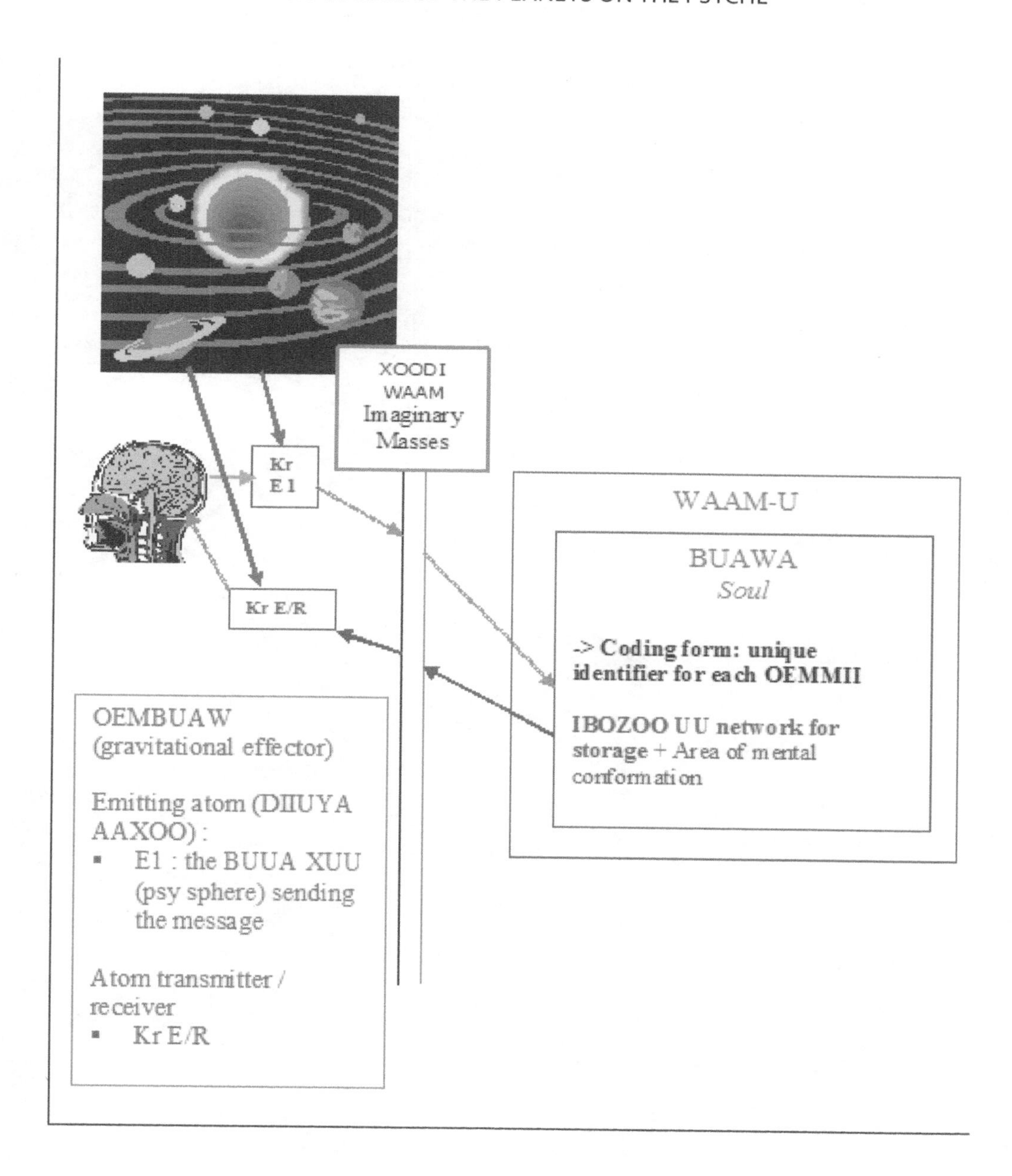

Initial gravitational flow
Gravitational flow regulator

Initial BUAWA flow
BUAWA regulated flow

TELEPATHIC COMMUNICATION

The evolution of living beings necessarily leads them to connect to their global-BB, to BUAWA and develop various forms of communication.

Oral communication of human beings seems to be generally followed by the development of various forms of telepathic communication. The principle of telepathic communication seems to be broadly the same for all OEMMII but physiological differences do not always allow inter-species communication.

The Oommo documents globally meet us on the operation of telepathic communications. But is telepathic communication specifically related to each BB? How can it be encoded? Are there any distance limitations for this type of communication? What are the factors that impact telepathic communication?

The telepathy is known and partly acknowledged by the scientific community of psychic sciences. Psychological approaches based on the holographic model in the sense of Karl Pribram have been attempted to explain telepathy based on the principle of non-locality. But without success and David Bohm aware of the complexity of the phenomenon cleverly thought that quantum non-locality was insufficient to explain telepathy...

Let's see information that our Oommo friends have sent us...

CONTEXT

Letter: 337 BUUAWE BIAEI DISCOVERY

"—For the first time we checked that codified movement of electrons in such atoms were exactly like telepathic transmission.

—We discover, contrary to what was believed, that the telepathic transmission is received simultaneously by all human beings; although a subconscious mechanism takes care of blocking it, that is to say, to prevent the passage of a message to people to whom it is not intended. "

"If the telepathic transmission requires pairing pathways in humans (nerve pathway) to move from one soul to another, it is because the collective soul and the individual soul are two independent entities linked only by the body as it is alive."

"You can observe if the telepathic transmission was realized exclusively in the BUAWEE BIAEI (COLLECTIVE SPIRIT), the total time of transmission would be ZERO because the human brain absolutely not intervene."

"The process for communicating with people that I know the identity... It occurs in this brain area two types of pulses: a bit like if it were two transmitters. If one of you is a technician in television, he better understood when compared with the signals you call IMAGE and SYNCHRONISM.

The first ones, properly codified, would be nerve impulses we call BUUAWE BIEE, each during 0.0001385 second. They are a kind of telegraphic message which is sent through one of the Krypton's atom we call BUUA XUU.

The second group of signals, which are also codified, is characterized by the fact that time has longer pulses: 0.006385 seconds.

What is the function of the first transmitted signal? Carry out easily the verbal content of our message, words that express the ideas we want to reach our interlocutor located at a great distance from us. But this message is received by all men of UMMO (we repeat that cases of telepathy recorded by you confirm that the phenomenon is identical for humans of the Earth).

How it is that one person is able to interpret such a message and be aware that it is directed only to him? The second group of pulses gives us the answer.

The code used to identify a person does not differ much between us and you in real life. Thus, as the name UGAA 4, son of YODEE 347, is mentally associated with a particular individual, a sequence of binary numbers, sent as pulses, will be used to specifically distinguish the person receiving the message.

Imagine a million safes located throughout your country. Each can only be opened only by using a combination of six numbers and all under the supervision of guards.

Inside these boxes there is the key to the interpretation of any coded message.

Now, send encrypted letters that can only be read with the help of cryptographic keys locked in coffers.

Send a million copies to all the guards with a single indication of the envelope for example 763,559.

Only the recipient who has the number of the coffer mechanism which coincides with this number can open it and know the content of the message.

The example is well chosen precisely because the selection of a person is obtained by a similar physiological mechanism to those coffers on earth (UMMO they do not exist) and a neural network establishes synaptic connections or disconnections based on an encrypted sequence of binary pulses which represent 1700 digits. "

D45 "BUUAWE BIEE TIME = 0.00013851 seconds; (time it takes man to pulse unit through the BUUA XUU (PSY SPHERE) to another man of UMMO by telepathic means. "

NR18 "The distance does not matter to establish the telepathic but interference from the presence of imaginary masses in the OUWAAM would slightly affect the synchronism when communicating with a very great distance."

ASSUMPTION

This is essentially a reading hypothesis including interpretations and extrapolations of the indications given by our visitors.

Telepathy thus functions roughly like "FM" with a carrier wave of the message associated with an identifier wave for the recipient. The Oommo people indicate that the telepathic impulses pass through BB, without specifying whether it is the global-BB or the planetary-BB.

The idea is that the encoding structure of telepathic messages is only related to the species of OEMMII. It would be totally independent of the planetary-BB of the OEMMII.

The Oommo people indicate that the telepathic impulses pass through BB, without specifying whether it is the global-BB or the planetary-BB.

The idea is that the encoding structure of telepathic messages is only related to the species of OEMMII. It would be totally independent of the planetary-BB of the OEMMII.

The scope of telepathic communication is independent of the distance, because it transits through the WAAM XOODII to reach the cosmos WAAM-UU with all its planetaries-BB. Only interference due to the presence of imaginary matter in UWAAM would know slightly affect the synchronous communications at a very great distance, taking account of gravitational effects that may disturb the telepathic impulses.

In principle, all OEMMII of the WAAM-WAAM should receive telepathic emissions of all OEMMII.

Generally, the precise structure of OEMBUAW of each OEMMII species should selectively capture specific frequencies to its kind.

DESCRIPTION OF TELEPATHY FLOW

Telepathy therefore operates in "frequency modulation" with a carrier frequency of the message and an identifying frequency. The transmitter sends the message and all receivers receive it. In associated manner, the transmitter sends in "parallel", the ID code of the message recipient. The recipient of the ID code probably corresponds to a structurally specific mental picture.

All OEMMII receive the message and an identifier that does not correspond to them, in this case the message is not processed by the OEMMII receiver.

The recipient receives the identifier corresponding him and allows the passage of the message associated with its brain, which decodes it, then transfers following the normal BUAWA process. The OEMBUAW converts telepathic signals in the brain by biochemical signals that generate mental images intelligible for the brain, which are transmitted to BUAWA.

The pulses which encode the recipient, encode the mental image associated with this recipient. Mentally evoke his name may be sufficient to generate the recipient's mental image. This mental image is encoded and transmitted in the WAAM-UU.

Message encoding

Mental images telepathically transmitted are encoded in a universal form. They encode language or perceptions.

This does not mean that there is a universal language, but simply that for any OEMMII emitting a coded mental image "A", then any OEMMII OEMBUAW will capture a coded mental image "A".

Then, everything will depend on what the recipient's brain will do with this mental image, if the mental image makes sense or not. This would explain that some OEMMII can "read" our thoughts and learn our languages by directly decoding the pulses emitted by the OEMBUAW of our brain, either:

a priori, the krypton atom transmitter of telepathic messages (E1 on synthetic scheme at the end of chapter)

possibly, the krypton atom transmitter of messages to BB and BUAWA (E/R on synthetic scheme at the end of chapter)

Conversely, some OEMMII could issue direct messages in the form of telepathic impulses to the receiver atom UAXOO of OEMBUAW.

This would explain, so that these can technically issue OEMMII telepathic messages directly to terrestrial humans.

It is customary for exocivilizations who visit us, to learn and use the languages of the visited planet. Consequently, there is nothing surprising that a terrestrial receives a telepathic flow in its own language.

In our brain, there is no difference between the mental representation of a vocalized word or a word heard in a dream for example. This activates in the same way the same brain areas. Telepathic words transferred to the person will be decoded by the brain in the same way that a vocalized word.

The effectors of GESI2C3H3

We mentioned in *chapter information flows of living species,* the hypothesis of a relay-compound of GeSi2C3H3 that would do the interface between Kr2 pairs and DNA. It may synchronize all the cellular elements needed to drive mutations led by the planetary-BB.

It was the one who would protect DNA against undesirable changes and also trigger controlled mutations.

In addition this compound—Germanium, Silicon, Carbon, and Hydrogen—would be the interface between the bio-frequencies from neurological processes and different crystals and rare gas, like krypton, located in the brain.

The carbon atoms and hydrogen are the interfaces between the nerve brain impulses and telepathic communication.

D41: The localization of these krypton atoms in the human body is very difficult for the following reasons:

–THEY ARE NOT COMBINED WITH THE REST OF ORGANIC OEMII COMPOUNDS (human body).

–THEIR NUMBER IS VERY REDUCED (we've counted 16)

Some are located in the TEMPORAL LOBE of THALAMUS in the HYPOTHALAMUS and other areas of the BRAIN BARK.

Schematic diagram of neural-telepathy stream

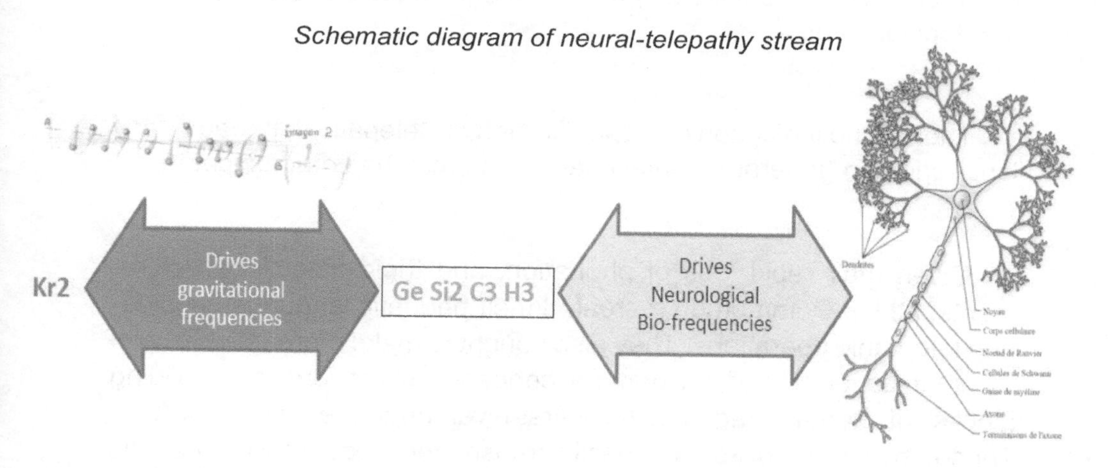

D45: These are the nerve impulses which, thanks to the different carbon atoms and helium whose QUANTUM statements have been excited, modify by resonance, the ordinary states of zero frequency (plane wave) of each atom of Krypton by OWEEU OMWAA effect. Thus the memory message, for example, will be encoded into these atoms with this shaped WAVES.

"Nerve impulses we call BUUAWE BIEE, each one during 0.0001385 seconds. They are a kind of telegraphic message which is sent by one of the atoms of Krypton"

So the encoding bio-frequency of neuro-telepathic stream is 1/0.00013851 is just under 7,220 KHz.

A periodic function square waveform: this is what captures our BUUAWAA (soul).

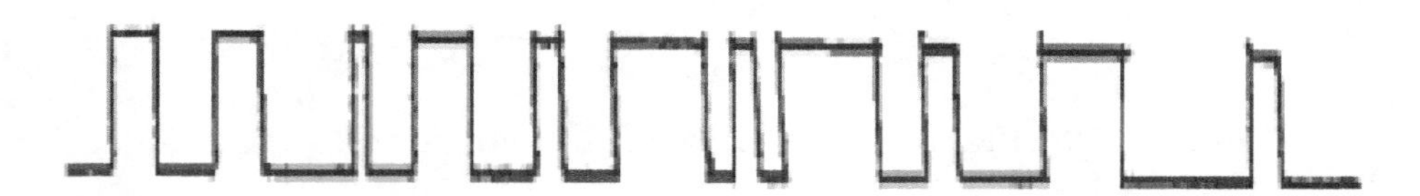

OOMMO TELEPATHY

Regarding the Oommomen, the "shape of OANNEAOIYOYOO communication", the telepathy transmits messages comprising "simple and topical ideas".

There is no indication of a specific kind of "telepathy language", any language can generate mental images and may be communicated.

Given their rapid loss of phonation, and their dialectical requirements, I think Oommomen recreated their phonetic language considering telepathic operations. They also sought to make more reliable, with communication based on primary concepts, which are the "building blocks" of mental images of their first-level language DU-OI-OIYOO. These "building blocks" are easily transmitted sequentially phonetic unequivocal.

The origin and telepathy learning for our UMMO friends:

> Telepathy is a natural step in the evolutionary process resulting from the continued increase in cortical mass and complexity of synaptic connections and neuronal glial. The cortical architecture specific to telepathy, among OYAGAAOEMMII exists at a primary level at neuronal glial groups, specific to language linking the angular gyrus and Broca's area.
>
> However, the complexity required for full functionality is reached only to the restricted part of

the population of OYAGAA having undergone mutation M130 (haplogroup C DNA of the Y chromosome).

The development of these neuronal glial groups remains inadequate in most other racial groups, particularly in individuals with a predominant haplogroup R-type (white people), except in cases of favorable individual mutations.

The OEMMII of OYAGAA also has the disadvantage of the one-sidedness of brain areas dedicated to language. Limited capacity of telepathy remains possible, with a rigorous exercise, with the plasticity of these brain areas and multimodal nature of neurons involved in the process.

Telepathic ability develops, dice the beginning of adolescence, practicing a consistent exercise that requires a lot of calm, concentration and intimacy with an emotionally very close relative.

This is a learning process that psychodynamic on OOMMO, requires several terrestrial months to get a consistently reproducible connection, progressively strengthening the synaptic circuit effectors.

This is for the single parent with whom the child is exercised during this first phase. The process continues, faster each time, with all members of the immediate family.

Telepathic ability of children tested for orientation, 40 XII (OUMMO days) before integration to OUNAWO OUI (university) at the age of 64.67 XEE (13.7 years). This capability will continue to expand, for emulation and affinities between the young people who come together every day.

We can also note that oxytocin, a peptide hormone synthesized by the supraoptic and paraventricular nuclei of the hypothalamus and secreted by the posterior pituitary (neurohypophysis), promotes learning telepathy.

The remote viewing

The "Remote viewing" is probably antediluvian and is experienced in the 1930s by JB Rhine, Upton Sinclair and René Warcollier. This work is continuing in the 1970s with the "American Society for Psychical Research" and "SRI International", with the work of the physicist Hal Puthoff. Intelligence agencies of the main countries have developed this research, sometimes officially as the CIA and DIA (Defense Intelligence Agency) with the "Star Gate" program.

The telepathic nature of communications is explained by the cosmological framework. The RemoteViewer connects to the targeted person, or retrieves a connection without knowing the person. This person serves to be a "camera", and transmits information which is felt, more or less clearly, by the Remote Viewer.

Remote care

Remote care practice is close to the remote viewing. We saw that the aura and body are strongly interacting. We believe that remote care practice is also using the telepathic channel. This, in several possible ways. Either affecting secondarily the aura of the patient or directly on the patient's subconscious, so that himself causes a psychosomatic effect.

Conclusion

The neurological evolution of Man to access to opportunities for telepathic communication is perhaps not as distant as one might think at first.

Indeed, it is a fact today that neural structures can be reconfigured dynamically by sustainable "mental gymnastics" practices commonly called "meditation". This has been shown by several types of experiments, in particular by Mathieu Ricard and Richard J. Davidson. This gives us opportunities to use these skills...

The synthetic scheme of telepathic flow:

Soul emitter—> krypton—> Brain—> krypton—> WAAM-UU—> krypton—> recipients brains

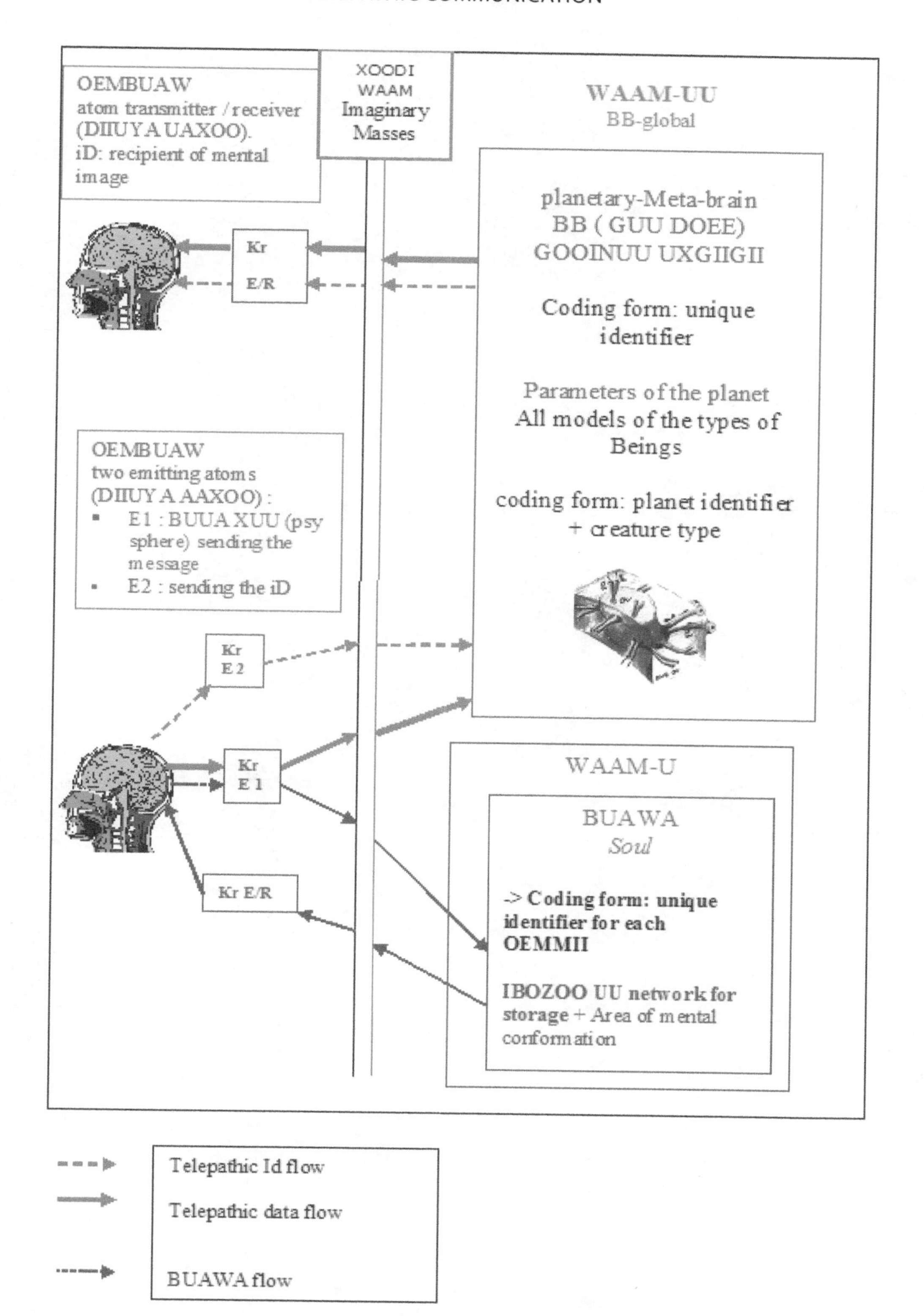
OEMBUAW
atom transmitter / receiver
(DIIUYA UAXOO).
iD: recipient of mental
image
XOODI
WAAM
Imaginary
Masses
WAAM-UU
BB-global
planetary-Meta-brain
BB (GUU DOEE)
GOOINUU UXGIIGII
Coding form: unique
identifier
Parameters of the planet
All models of the types of
Beings
coding form: planet identifier
+ creature type
Kr
E/R
OEMBUAW
two emitting atoms
(DIIUY A AAXOO) :
E1 : BUUA XUU (psy
sphere) sending the
message
E2 : sending the iD
Kr
E 2
Kr
E 1
Kr E/R
WAAM-U
BUAWA
Soul
-> Coding form: unique
identifier for each
OEMMII
IBOZOO UU network for
storage + Area of mental
conformation
Telepathic Id flow
Telepathic data flow
BUAWA flow

COMMUNICATION WITH "SPIRITS"

The OUI-JA, turning tables are known to be spiritualist practices. The few scientific experiments conclude that "the subject performs unconscious muscle movements" called ideomotor effect. Which is perfectly correct, but does not explain the causes of these unconscious movements or why the unconscious ideas generate precise and very concrete actions...?

The assumption is that a telepathic flow, conscious or not, of sufficient levels, cannot target a living human, but the meta-brain planetary-BB itself and his integrated souls-BUAWA. This would create a feedback, conscious or not, able to be transposed by human receptors via a OUI-JA game, for example.

> U. — Nous sommes en contact avec le B.B. par le U. —We are in contact with the BB through our subconscious. Under these conditions the communication is possible, but in practice it is difficult to know it (to check it).
>
> The psychic phenomena of communication are not always real, but they can be. It follows that it is unclear whether they have passed through the telepathic channel, and we may be wrong.

D731: We know that when our death will occur a merging, integration, a close liaison of the psyche, of our "spirit" (nor hardware, nor intangible but all the information matrix of our life) with the "universal" collective psyche.

We can connect us more with loved ones, communicate with the spirits of the other deceased brothers, participate in the global knowledge of all the biosphere, not only OEMII who just died, with all humans but even since born life on OYAAUMMO (and, of course, for you, from Homo habilis to the last of your brothers).

Is also possible knowledge of the real world including living beings, since BB is informed of the whole process of living beings that are not yet dead.

This means that the deceased OEMMII through his psyche can somehow influence his most loved ones through the unconscious and, to some degree also the things around them, since the biosphere changes the surrounding physical environment through living beings.

B. B. is the collective Mind. We can also call it subconscious or unconscious collective, since its contents are operational but are not made aware of us other ALIVE BEINGS.

The mind of a deceased brother can be, and in fact sometimes does, we attend, we protect and sometimes interacting so VERY ACTIVE...

—A Soul in the BB can she see the future of the Earth?

U. —NO.

Access to information structures of meta-brain planetary-BB may explain why OUI-JA participants can sometimes express information that was unknown to them.

Someone sent information to planetary-BB, it exists and is potentially accessible. By cons, no chance to try to see into the information for the next lotto numbers, because the information does not exist!

As we have seen in previous chapters, infinitesimal quantities of material can be sublimated into huge amounts of energy, they are LEIYO effects of dimensional axes switches.

If we push the reasoning of our hypothesis, we can imagine that the flow from BB can create micro-dimensional effect switches, to impact onto the matter of our three-dimensional framework.

In other words, the flow from BB relayed by receiving human antennas could create micro-dimensional effects of dimensional axes switches to produce different effects according to their angular torsion.

As we mentioned in subchapter "The dimensional axis switches" in the "A new cosmological paradigm" of these dimensional axes switches can be total or partial. The axes of the masses and volumes can be "twisted" and thus produce a mass loss or disappearance of volumes, and often both at a time...

We can note that by heating a OUIJA stool, it accumulates thermal energy that excites the electronic layers of the atoms and makes them more susceptible to bio-frequencies, it follows that the LEIYO effect of the projection on the stool is faster than without heating...

So we'll light the mysterious phenomena "paranormal activities" with the cosmological model and his practical implications. We will present the different scenarios that can rationalize these phenomena.

Pre-cognition or vision of the future

The concepts of precognition or mediumistic vision of the future are linked to so-called phenomena of "prophetic dreams" or psychic visions.

Does actual future pre-exist?

We saw the potentialities of the Absolute Reality partially become actual. The Absolute Reality, AIIODI, interpreted by humanity is constantly created and dynamically, while all the potentialities is an almost infinite tank but static. So we can change the course of our real by thinking differently our future...

Actual future therefore does not pre-exist. Note that if the future pre-exist, the present or the past would be meaningless, time would not exist neither and everything would be static.

However should we consider the millions of people who have "prophetic dreams" everyday result of mere chance? Just as the psychic forecasts?

Note that subjects under hypnosis increase their psychic abilities and in particular in terms of precognition. Furthermore, studies by Louisa Rhine revealed that the precognitive visions are—mostly—tragic premonitions of unfortunate events being four times more numerous than those of fortunate events.

Note also, that people in cultures where there is a practical Shamanism had realized better extrasensory perception tests.

The researchers on this subject are very numerous, it would be difficult to mention them all, as the writer John White, parapsychologist Stanley Krippner, physicists Harold Puthoff and the Russel Targ, psychiatrist David Loye, etc. But are very poor and in the absence of a satisfactory theory, they refer to the holographic theory developed by Bohm and Pribram. This theory making no concrete answer on this issue...

Let's start with a concrete case, which was exposed by a friend, Marc. This boy had spent more than a year in a hospital bed for spinal problems had time to develop skills for meditation. A few months after his convalescence Marc had a dream or rather a nightmare that woke him. He was driving on a small road in the south of France and out of a turn, uncorked a vehicle on the right to cross the road by cutting his way. The collision occurred, causing his death. The dream was of a supported realism. A crashed white car, a Peugeot 206, the driver, a man in his 70s, in plain clothes, beige, white... But a simple bad dream soon forgotten...

About a week later, Marc takes his car to get to an appointment. This small Provencal's road that leads to his appointment reminds him of something, but what? Then he gets it! The nightmare of the past week! Yes, it recognizes the turn! Marc brakes immediately. He is quite right. A white Peugeot 206 cut off the road to cross the other way. Marc honks to the driver who parks his car a little further. Marc joins him, he's a man in 70 years in plain clothes ... he is so sorry...

To resume a famous French actor, Louis Jouvet's tirade in the movie "Drôle de drame": It's weird, so weird...

Here is my interpretation of this classical case, which is generalizable to all cases of this type.

We know that the meta-brain BB is informed by the psyche of the "man in white" that he plans to take his car, the white Peugeot 206.

This, either by routine or because the man thought about it previously. "Oh! I miss this nice rosé de Provence, I will buy it next week at my favorite winemaker..."

It is the same for Marc and for all people for the event. The meta-brain BB is informed by human psyches of their intentions before they do make them. This sometimes very long before...

It is easy for the meta-brain BB to predict what will happen. The collision prediction is "calculated" by the meta-brain BB. The meta-brain BB does not want to lose those two "good customers" that go back to him full friendly information on Provence, and certainly have not finished their job on planet Earth... Then, meta-brain BB sends an alert message. Probably to all mankind, but only the subconscious of those concerned will take into account. Marc in particular will be awakened by this nightmare, allowing him to remember...

This is the "calculation" power of meta-brain BB that allows it to anticipate future situations. Individually each of the human actors of the event has only a fraction of the information and in no way can make such forecasts.

Thus, when they are ultimately faced with the event, it appears that their dream was "prescient" ...

Similarly, some mediums can receive this type of information from the meta-brain BB, a priori, this with an anticipation that is not limited in time. The only limits to the predictive anticipation are the limits of the "calculation" power of the meta-brain BB.

In the 1960s, Marie-Louise Aucher (1908–1994), musician and singer, discovered vibrational correspondence between sounds and the human body: each sound corresponds to a part of the skeleton. It is understood that the human body has a transmitter/receiver function when a sound is emitted from the resonant frequency of some parts of the skeleton.

An instrument producing a given frequency can resonate with a body part and reciprocally the human voice can resonate with an instrument. That is psychophonie. Many resonance phenomena exist and their effects are sometimes mysterious and surprising...

The transcommunication presents an analogy with psychophonie in sound frequency, the difference being the range and nature of the frequencies of the transcommunication. Friedrich Jürgenson incidentally discovered the phenomenon in 1959 and Senkowski Ernst, professor of physics, invented the concept of "transcommunication"

Raymond Bayless, Attila Szalay and Konstantin Raudive pioneered experimentation on this topic. Pr. Marcello Bacci plays with the tuning key of a radio receiver tubes between 7 and 9 MHz. Among the messages he would have heard "POWER OF THE SPIRIT, opportunely modulated IS TRANSFERRED TO THE PERSON RECEIVING which PROCESS WITH HIS BRAIN ELECTROMAGNETIC WAVES IN GIVING THE RADIO TRANSMISSION."

Friedrich Jürgenson has heard a message relating to a medium wave frequency 1480 kHz and Hans Otto König two information about short wave frequencies of 10 MHz and 7 MHz.

U. —The psychophonies [the transcommunication] are produced by electromagnetic radiation. It is a PHYSICAL phenomenon because the recorder may, under specified conditions, be converted into a receiver.

An induction coil connected to a capacitor form an oscillating circuit capable of receiving the electromagnetic waves from a transmitter. In the transmitter under certain conditions, a resonance can happen in the resonant circuit.

If another oscillator circuit has the same characteristics ... that is to say that the induction has the same number of milli henrys, the capacity of the same number of faradays, or combination thereof, therefore the capacity of the induction will have the same characteristics ... under these conditions the transmission and reception of electromagnetic frequencies is possible because the two circuits are in resonance.

Into a tape recorder sometimes occurs resonance effects with electromagnetic emission whose frequency is received. Although the tape recorder was not designed to capture electromagnetic waves, however, it is able to pick them up. It is in these conditions that we receive the psychophonie.

An element capable of emitting electromagnetic waves may also issue a PSYCHOPHONIE ... and be picked up by a tape recorder...

The "majority" of psychophonies [Transcommunications] are captured through electromagnetic oscillations.

The instrumental transcommunication operates on resonances between electronic devices and inductive electromagnetic devices or resonances with frequencies emitted by the human brain activity.

The unconscious telepathic communication is transformed into electromagnetic frequency by the person who is sending in this frequency range, and if the inductive device resonates, the frequency is normally processed by the camera to produce the sound or recording.

This may resonate with the tape recording head, certain components of the amplification circuit of the radio station or the windings of the speaker himself...

Schematic flow of OUIJA

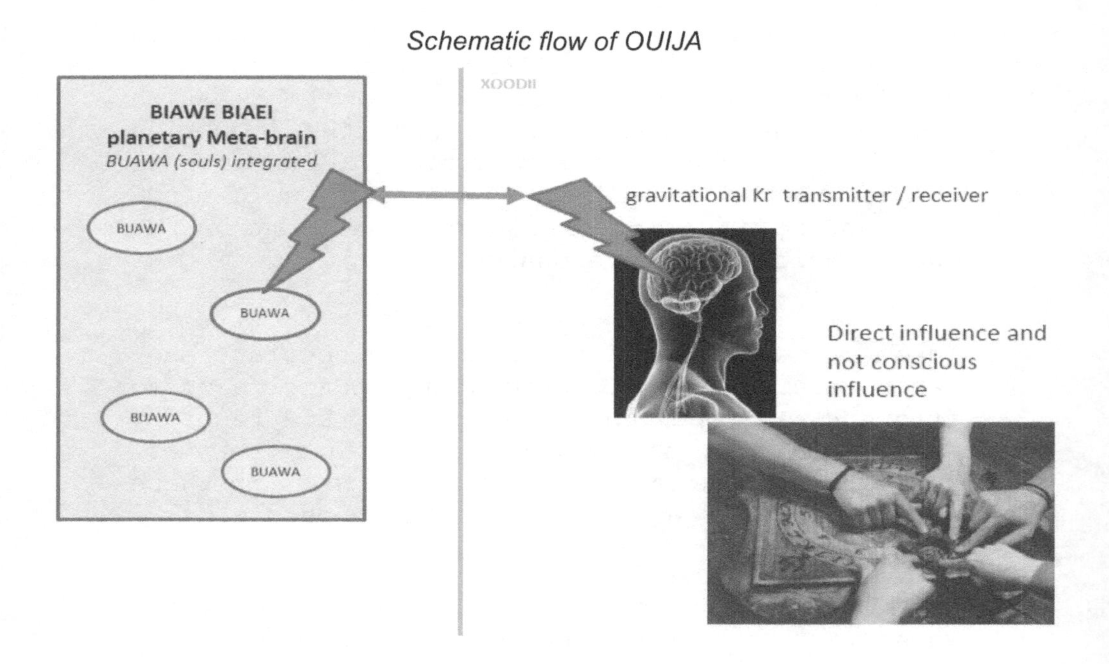

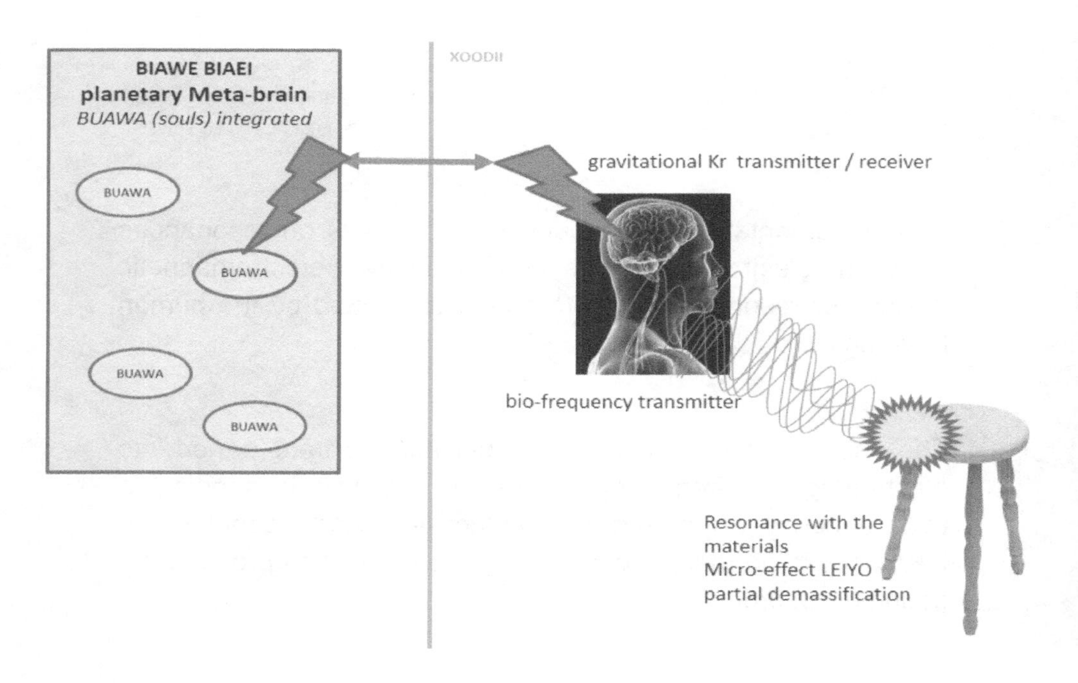

Schematic flow of Instrumental Transcommunication

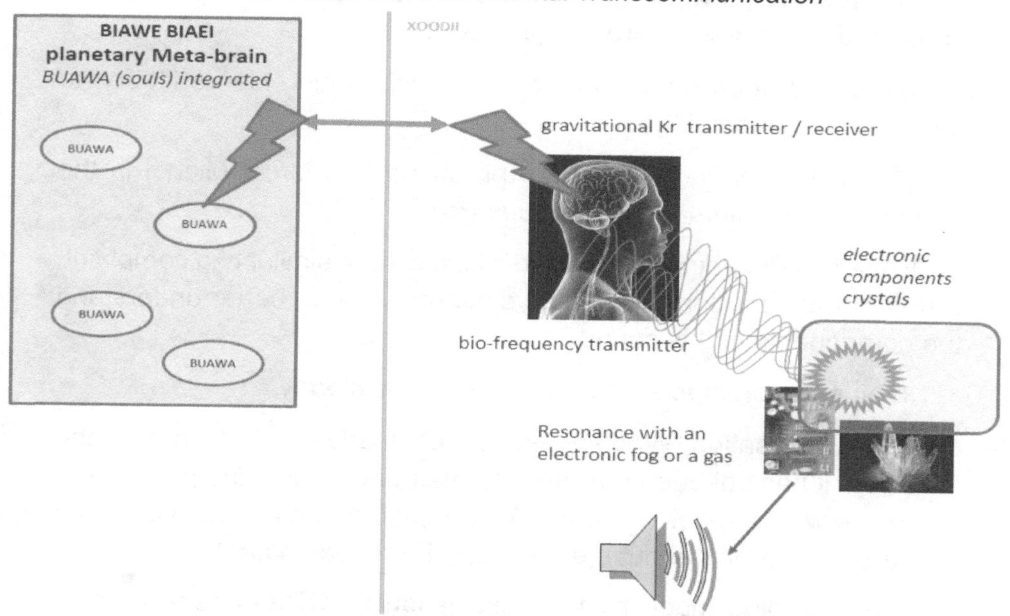

Heating a OUIJA stool, it accumulates thermal energy which excites the electronic layers and makes them more susceptible to bio-frequencies, it follows that the LEIYO effect of the projection on the stool is faster than without heating.

HEAR "VOICE" OF BEYOND

The stamp of "voice" of the hereafter is far different from that of humans. In the 2000s, measurements of recordings made by Michel Marcel for CERPI (CENTRE FOR STUDIES AND RESEARCH ON Unexplained Phenomena) indicate that they are about 16 times faster than the propagation of sound in air. This has the effect of being "voice" highly deformed relative to the voice produced by a set of vocal cords, which unambiguously distinguishes them from human voices.

To be truly heard and understood by everyone, they require signal processing. The "voice" recorded does not include fundamental frequency, but only harmonics and in the low-frequency sound (10 Hz to 20Khz) almost always low.

The association CROPS and Michel Marcel have kindly sent us several records with sounds that had not been heard by those present.

The recordings were performed either on magnetic tape medium converted into digital or direct digital records.

We called upon the expertise of a musicologist, Hervé Noury, to analyze these recordings.

The first recording series was produced by Marcel Michel in the presence of "medium" Jean-Jacques Poirier.

In the first recording, we spotted a kind of roar, similar to a complaint. Different transpositions and replays did not allow to better understand the components of this sound.

In other recordings are heard some words clearly.

The conversation "Who speaks to me" clearly hear a ' It's me ' and "Yes", but it has not been possible to better understand the passage or that voice would speak its name. Also a good means ' Yes I Can ' in a conversation talking about the creation of a research center.

The recording made by the association CROPS "I cannot do it" explicitly answered a question of the 'medium' Monique AUBERGIER.

In all cases, the tonal inflections of sound as well as permanent alteration of the formants [the energy maxima of the sound spectrum of the sound of speech] are those that usually produce a voice.

Note that in any case, these voices have not been changed by acoustic means (box, auditory mask, wall...) or electronic (filters, vocoders, saturation...).

A listening, these voices sound like the human voice had been treated using a vocoder, modulating white noise (continuous whooshing sound) but retaining tonal inflection of origin.

For some reason, not discriminating, these sounds do not involve fundamental harmonics. It is not possible to know whether this is due to a technical element or if the source itself does not contain fundamental harmonics.

We have little consonants, but vowels, that confirms the sonogram analysis recordings. Low frequencies dominate, there are few acute frequency characteristics of consonants. So we observed with the sonogram that the acute frequencies are missing, hence the inability to understand consonants.

The electronic identification markers of «voices»

An important advance was made by my friend Thierry Wilson from TCI-France. A sound engineer by training, he analyzed thousands of voice recordings in transcommunication. He found that almost all of these recordings contained what he called «markers» at the beginning and end of the recording. These were compacted sequences containing the structure, the rhythm, the template, of the sequence of the vocal message.

But, what about the rare transcommunication voice recordings where there were no such markers?

During our exchanges, we noted, in fine, that these recordings concerned a precise context. The people who sent the transcommunication were not related to the deceased.

Only transcommunication transmitters in relation to a deceased produced recordings with the electronic identification markers of the «voices». This is the beginning of proof and systematic recognition to differentiate the origin and nature of the flows recorded.

The stream structure contains packed sequences before and after the voice message sequence. These markers determine the structural parameters necessary for the voice reproduction of the message. It is likely that the markers codify the sounds to be produced with: the frequency or tessitura; the intensity; the timbre which depends on the frequencies contained in a sound which are superimposed; the voice reproduction rate; prosody or intonation.

These structural parameters or identification markers (*) of the messages are always in line with the nature of the recording medium. That is to say that if the recording medium allows recording in such and such a frequency range, then the message will be reproduced in the appropriate frequency range. The structural parameters of the messages intelligently take into account the nature of the recording medium for which it is intended.

interpretation within the cosmological framework

In all cases we were able to categorically eliminate animal cries or noise signatures. All recorded sounds are composed of formants characteristic of a human voice, although these formants were not perceived in the sound field by those present. The electronic identification markers of the «voices» prove that the nature of the transcommunication voices is very specific to the nature of the origin of the flow.

Let us study the causes of this phenomenon... The flow of these data from beyond space is done through cosmoses of a totally different nature. Indeed, these voices are above all data stored in the Soul-BUAWA of the deceased which has been integrated into the planetary Meta-Brain. The flow of this Soul-BUAWA, monodimensional and timeless, transits in the penta-dimensional space of infinite photonic speed which is the planetary Meta-Brain BB. It is he who actually takes charge of the flow and manages it, and not the BUAWA-integrated of the deceased itself. The integrated Soul-BUAWA thus establishes the same type of telepathic communication as for a human-to-human connection through the planetary Meta-Brain BB. We have seen the flow of gravitational frequencies coming in at the krypton atoms of the OEMBUAWE and coming out in bio-frequencies of 7.2 KHz on the Human side. The Germanium-Silicon compound would play the role of flux capacitor-transformer.

The difference in the nature of the streams and the serial transformations explain why there may be a difference of a factor of 16 between the speed of a human voice and the final recorded output streams of the deceased. We also note the absence of fundamental frequency of the BUAWA flow through the planetary Meta-Brain BB.

The Soul-BUAWA being timeless, as we have seen, it is made to transfer information by following the course of time. This is very different from a stream of words or images. This link does not occur between 2 brains of the same type which can exchange mental images coded in the same format. The Soul-BUAWA data format is a train of impulses for subconscious brain processes, but not a mental representation directly encoding sound or image intelligible directly by the brain areas of the terrestrial human. The earthly human is not wired for this, the BUAWA flow remains subconscious. To become intelligible, it must be transformed into frequencies that can be picked up by our neuroreceptor organs. The human bio-frequencies can serve as frequency modulating carriers of the flow of the BUAWA of the deceased which communicates through the planetary Meta-Brain BB. These bio-frequencies in modulation will be able to be captured with electronic receivers, crystals, etc.

When the integrated Soul-BUAWA establishes telepathic communication for a human through the planetary Meta-Brain BB, the one intelligently structures the flow so that it can be rendered in an understandable human voice. This stream structure contains packed sequences before and after the voice message sequence. These markers determine the structural parameters necessary for the voice reproduction of the message. It is likely that the markers codify the sounds to be produced with: the frequency or tessitura; the intensity; the timbre which depends on the frequencies contained in a sound which are superimposed; the voice reproduction rate; prosody or intonation.

Thus the voices of the beyond are not human voices, but the expression of a flow of data ultimately transformed into sound frequencies, according to its identification markers which codify the sounds to be produced.

The fact that the structural parameters or identification markers (*) intelligently take into account the nature of the recording medium, confirms my idea that the planetary Meta-Brain BB is at the origin of the parameterization of the flow of messages .

It seems to us that the packets and frames of these flows have an architecture very similar to those used in computer networks, of which here is a hypothetical explanation (see also the diagrams of the Theoretical model of information packets in frequency format)

Hypothesis of analysis of a sound communication on Earth (TCI)

How could a «deceased», whose soul is integrated in the Meta-Brain BB, could communicate a sound message on Earth (TCI)?

When a «deceased» wishes to communicate a sound message on Earth in TCI, he sends sequences of messages corresponding to the words he wishes to communicate. These sequences of messages encode in binary each syllable of each word that the deceased wishes to pronounce. Each message corresponds to a word of one or more syllables.

Each message is itself structured with several packets.

A header packet itself describing the structure of the other message packets. Thus, from the 2nd packet to the penultimate, the syllables of the desired word are coded in binary.

An end packet homologous to the start packet terminates the packet.

The flow coming from the Meta-Brain BB is therefore a binary frame which evolves at an infinite speed. This binary frame is structured by sequences of messages, themselves composed of a start header, syllabic packets and an end-of-message packet for the end of the word. The start frame could begin with a syncword allowing the receiver to adapt to amplitude modulation, adjust for variations in signal speed, and for frame synchronization.

It is necessary that the binary message which leaves at the infinite speed of the Meta-Brain BB, be transformed into a message entering the human brain at the speed of light. It is the krypton atoms that perform the conversion by emitting gravitational waves.

As we saw in the Bio-frequencies section of the chapter INFORMATION FLOWS IN A LIVING SPECIES, the hypothesis is that the compound GeSi2C3H3 plays a role of modulator-demodulator in various wavelengths. In particular by transforming these gravitational waves into different frequencies necessary for the synchronization of intracellular organelles.

The complementary hypothesis is that the compound GeSi2C3H3 also plays a role in the decoding of binary messages coming from the Meta-Brain BB. The receiver-transmitter human brain is traversed by a multitude of radio-frequencies of all kinds, including radio-frequencies that have bounced off the ionosphere.

The idea is that the binary message coming from the Meta-Brain BB is transformed by the compound GeSi2C3H3, into an analog signal comprising the radio-frequencies of the syllables of the word of the «deceased» message.

The GeSi2C3H3 compound is therefore crossed by a multitude of radio-frequencies of all kinds, including radio-frequencies that have bounced off the ionosphere. In the GeSi2C3H3 compound, the header packet is decoded with its descriptors of the other syllabic packets of the message. To achieve this, the GeSi2C3H3 compound uses the radio-frequencies that pass through it naturally. It filters and lets through in the outgoing flow, the bits of radio-frequencies corresponding to these descriptors, in other words, it lets out in the flow the radio-frequencies corresponding to the syllables of the word that the «deceased» wishes to pronounce.

The bits of radio-frequency, permanently transiting in the human brain, in the GeSi2C3H3 compound, which correspond to the syllabic descriptors are thus assembled, packet by packet.

Thus, an analog radio-frequency frame is constituted. This new analog frame can then be projected by resonance-induction by the «medium» on different media, magnetic or otherwise. This will ultimately result in the recording of a composite voice frequency, resulting from the multiple radio-frequencies that made up the analog frame thus constituted.

The screening process and the parameterizations of the descriptors are an invention of the Meta-Brain BB, adapted to the terrestrial operational context. Another BB Meta-Brain with telepathic humans might have done it differently...

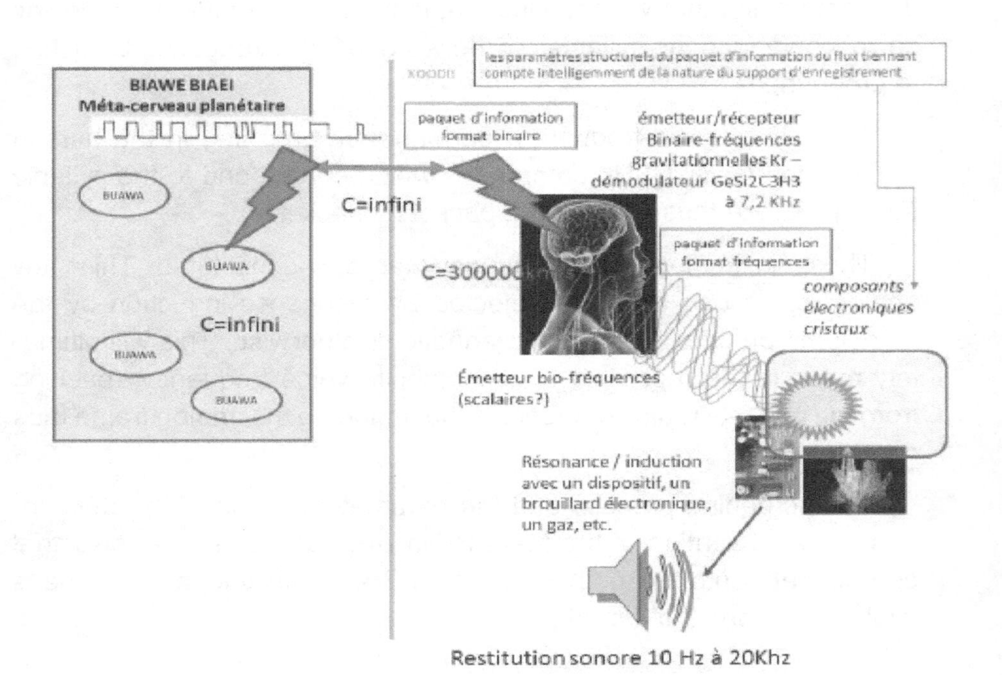

Modèle théorique de paquet d'informations au format fréquences

trame de début	1- Séquence sonore	2- Séquence sonore	3- Séquence sonore	trame de fin
Paramètres -1- Séquence sonore				Paramètres -1- Séquence sonore
Paramètres -2- Séquence sonore				Paramètres -2- Séquence sonore
Paramètres -3- Séquence sonore				Paramètres -3- Séquence sonore

La trame de début pourrait commencer par un syncword (en) permettant au receveur de s'adapter à la modulation d'amplitude, régler les variations de vitesse du signal, et pour la synchronisation de trame

MASS PROJECTIONS

As we have seen, bio-frequency projections can influence onto the masses axes to unmassify an object by resonance. But conversely, twists on the axes of the masses can generate mass, independently of the material itself.

On a given volume that can either be a gas or a solid object, the receiving person will project bio-frequency that will produce LEIYO micro-effect of a twist on the axes of the masses, creating in this case the mass onto any volume.

I was told by a friend "medium" case for which he had been asked. A lady had lost her husband and said that his ghost was still in the house. The lady said that often after lying down, the ghost of her husband came to join her in the bed. Although she could not see anything in the bed, it creaked and bent at the usual place of her husband. My friend "medium" did practices to boost the integration process of the residual aura of the gentleman and things again became normal.

Schematic projection flow of the mass

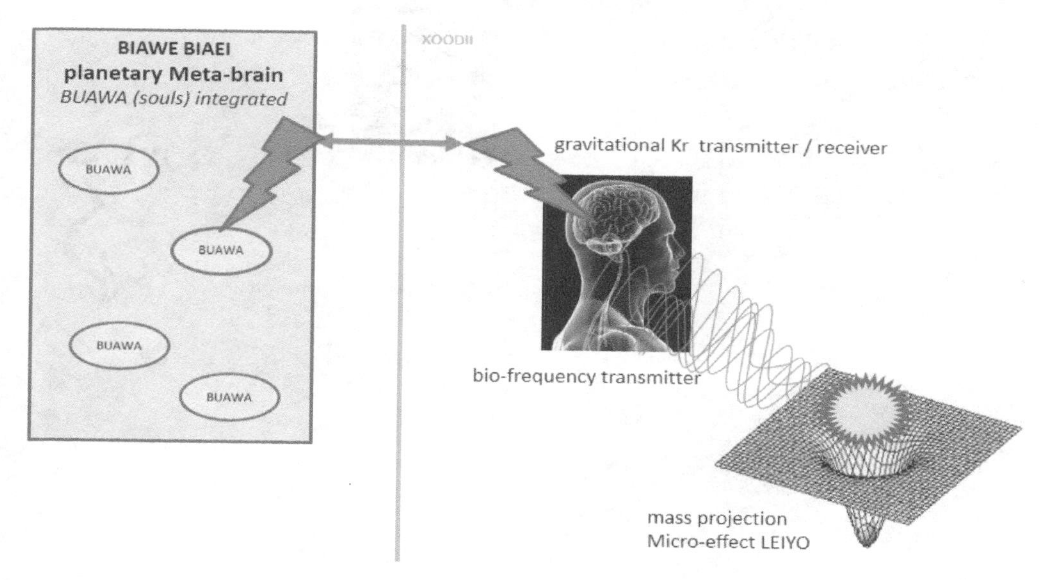

HOLOGRAPHIC PROJECTIONS

Similarly, a person will project bio-frequency onto an electronic device by resonance and electromagnetic induction, some bio-frequency resonances will produce effects on volumes, transferring in this case electromagnetic information, and creating a hologram.

These phenomena of resonant projections can be accumulated issuing the person ability. Thus, projections can overlay information on volumes and simultaneously produce or remove mass. This would allow the projection of what is often called "egregores", the 'mass ghosts "seen as living persons and can exchange a telepathic voice flows that will be experienced by the listener as an exchange of "normal" words. We will see similar phenomena in the chapter "*ABDUCTIONS AND MENTAL MANIPULATION*".

Diagram of volume flow or holographic projection

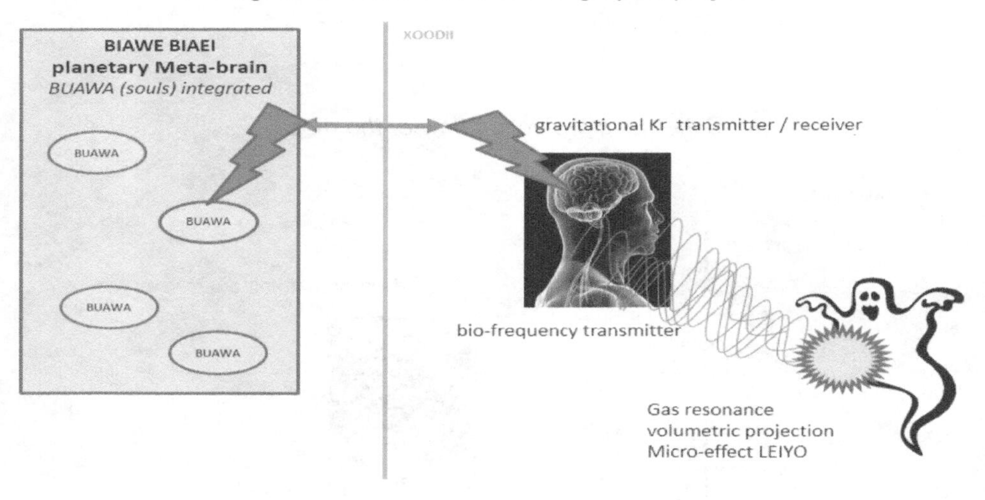

Volume or pattern of holographic projections and mass flows

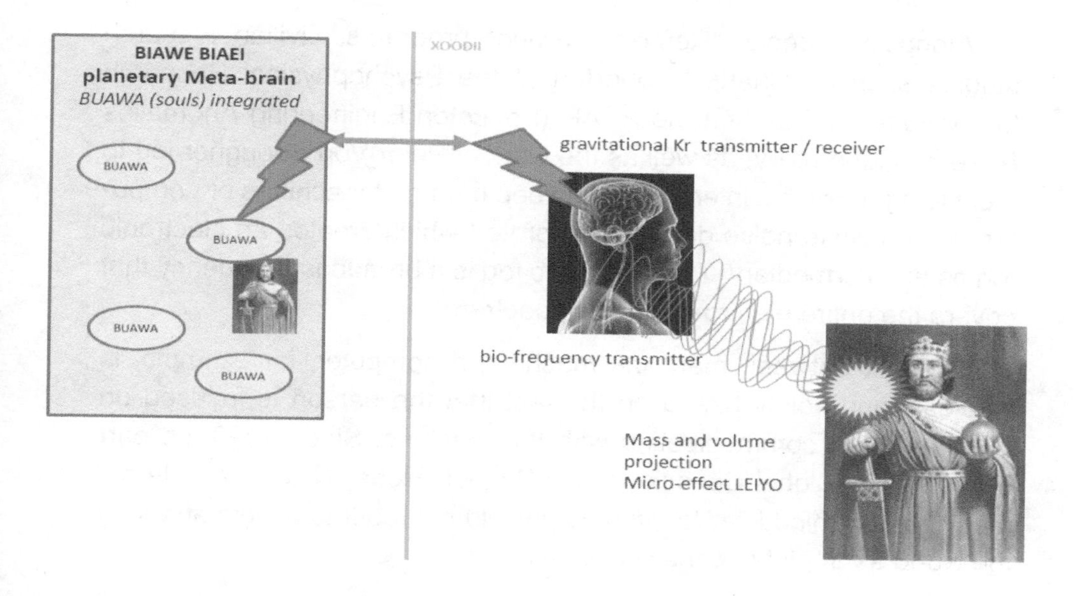

The Global Consciousness Project

Alongside special "Remote viewing" programs, civilian research studies such as Charles Honorton at the Psychophysical Research Laboratories (PRL) then the PEAR (Princeton Engineering Anomalies Research laboratory) as well as the work of Peter Von Buengner led to identifying a psychic interaction between man and machines or computers with a white noise diode incorporated which creates an electronic fog as an intermediary. The electronic fog is a broadcast frequency that covers the entire electro-magnetic spectrum.

The link between man and machine, a computer, for example, is wireless, and solely based on the fact that the person is focused on thinking about communicating with the machine. Since 1998, as part of the study "Global Consciousness Project" Roger Nelson wonders if many human minds, in "resonance", would not react to events affecting the world as a "global conscience".

Our assumptions on flows and projections can explain in detail the GCP. In the first example, the transition to the year 2000, an hour before the entire world population is feverishly preparing for the event. Connections to the planetary-BB are focused on this information, and culminating in the joy of New Year's Day.

Two streams can impact the stochastic white noise diode electronic device:

• either, the flow emitted by planetary-BB is sensed by the individual's subconscious near a computer of the GCP network. And then this person in turn emits a bio-frequency flow captured by the white noise diode electronic device. These are the individual's bio-frequencies that create resonance and break chance in the experimental *(FIGURE A)*.

• either, the flow of the planetary-BB directly acts on the stochastic electronic device, as for the relationship between the planetary-BB and OEMBUAWE, the human brain receptor *(Figure B)*.

Consider the example of earthquakes *(FIGURE B)*. These are perceived by various animals since the warning signs. The information is transmitted to the planetary-BB before the event is shown to humans. Again we see the break chance in the experimental device.

For these telepathic types of flows, the role of planetary-BB is to record and transfer information. Never planetary-BB modifies the contents of the stream. An alert launched by bacteria will be interpreted by the other bacteria and the planetary-BB. Although humans receive this bacteria-warning issue, it cannot be interpreted by them.

This leads us to conclude that it is directly the flow of the planetary-BB which is acting on the stochastic electronic device, as the link between the planetary-BB and OEM-BUAWE, the human brain receptor *(Figure B)*.

In the case of the World Trade Center, the planetary-BB was already alarmed at 4:00 pm so almost five hours before the impact of the first "airplane" (8:45) and half past six before the arrival of the second "airplane" (10:30). As the media had not mentioned the event before 8:45, the planetary-BB has responded to the stochastic electronic device when those who were aware of what was happening began to put their plan in action.

The planetary-BB is informed and can "calculate" events that will happen. Some "mediums" has "premonitory dreams" others have visions of the disaster before it actually occurs. This leads us to conclude that the planetary-BB has "calculated" itself the events that were going to happen and by itself alerted the subconscious human *(FIGURE C)*.

Schemes:

a) the case of New Year's Day 2000, flows pass through the OEMBUAWE (human brain transmitter/receiver)

b) the case of an earthquake flows pass through the BAAYIODUU (genomic animal's transmitter/receiver)

c) the case of the World Trade Center flows pass through the OEMBUAWE (human brain transmitter/receiver)

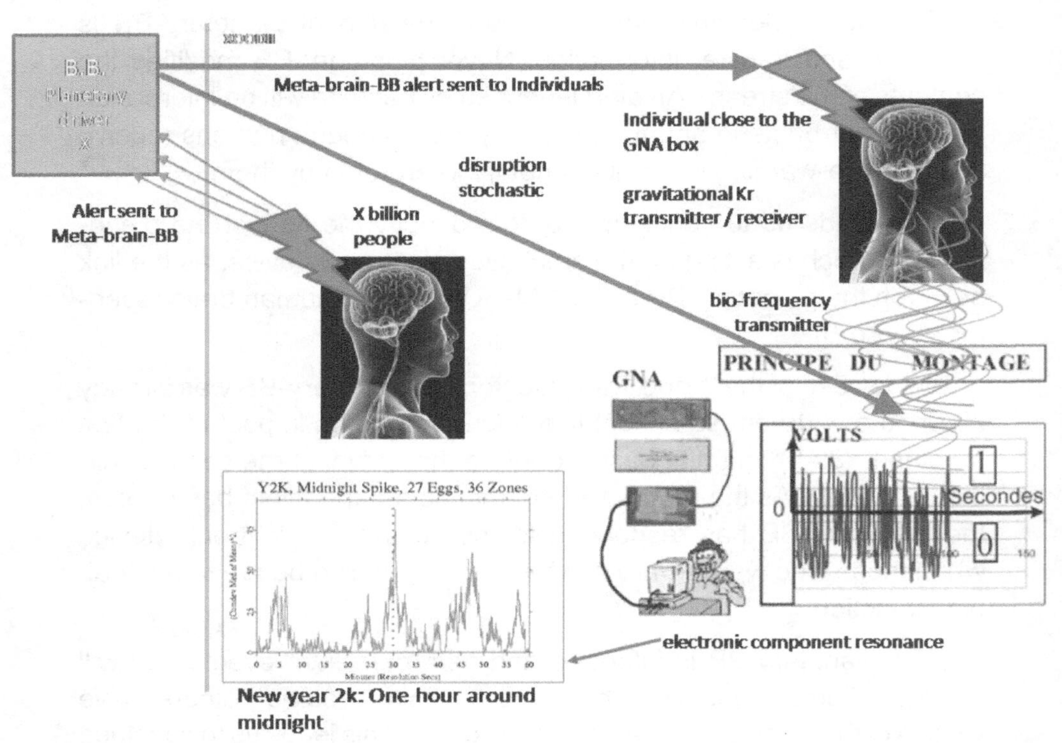
B.B.
Planetary driver
X
Meta-brain-BB alert sent to Individuals
Individual close to the GNA box
disruption stochastic
gravitational Kr transmitter / receiver
Alert sent to Meta-brain-BB
X billion people
bio-frequency transmitter
PRINCIPE DU MONTAGE
GNA
VOLTS
1
Secondes
0
0
Y2K, Midnight Spike, 27 Eggs, 36 Zones
electronic component resonance
New year 2k: One hour around midnight

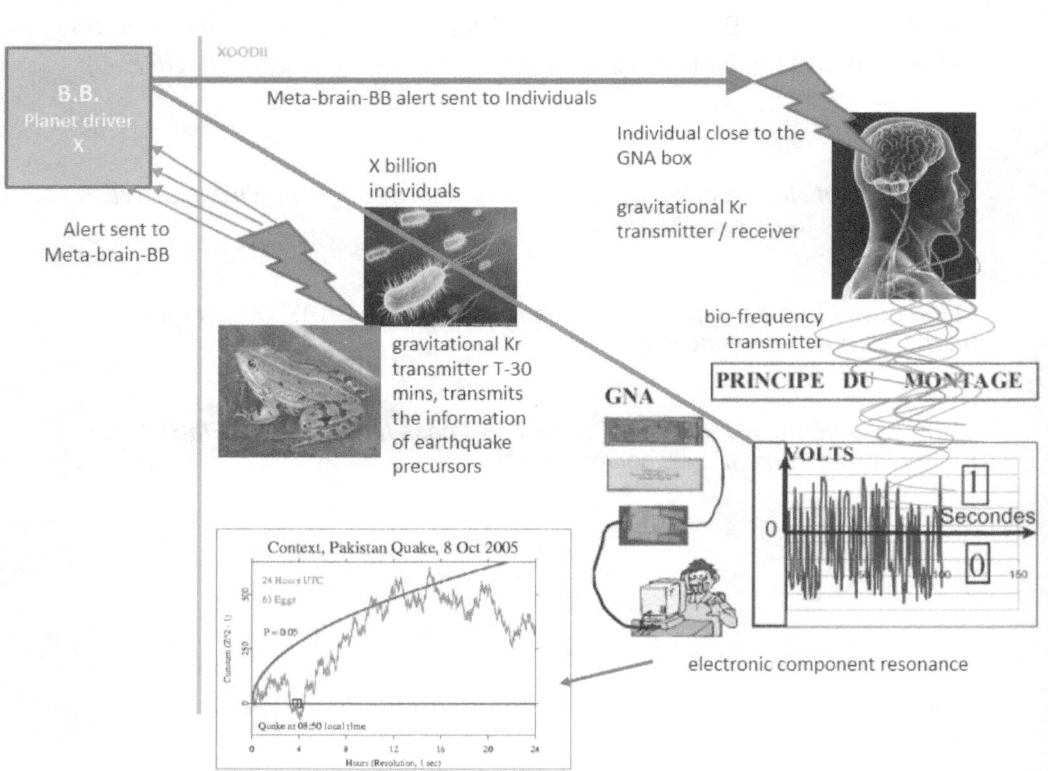
B.B.
Planet driver
X
Meta-brain-BB alert sent to Individuals
Individual close to the GNA box
X billion individuals
gravitational Kr transmitter / receiver
Alert sent to Meta-brain-BB
bio-frequency transmitter
gravitational Kr transmitter T-30 mins, transmits the information of earthquake precursors
PRINCIPE DU MONTAGE
GNA
VOLTS
1
Secondes
0
0
Context, Pakistan Quake, 8 Oct 2005
Quake at 08:50 local time
electronic component resonance

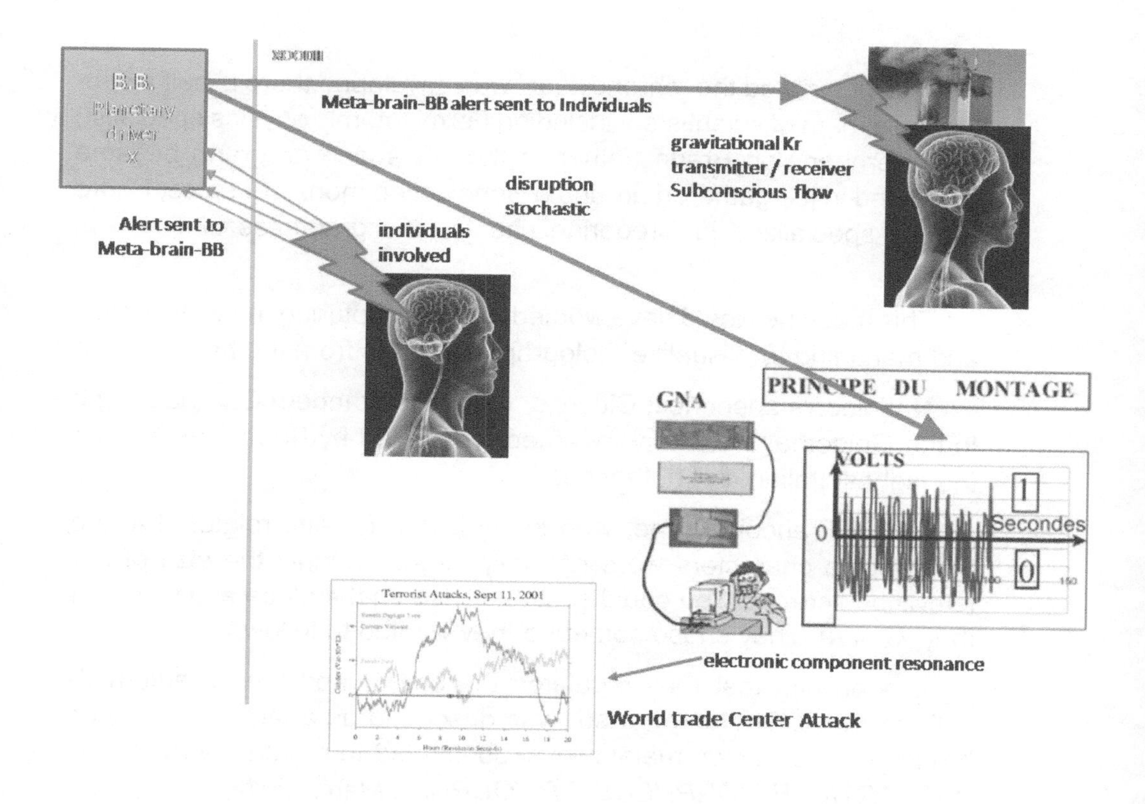
B.B.
Planetary
driver
X
Meta-brain-BB alert sent to Individuals
gravitational Kr
transmitter / receiver
Subconscious flow
disruption
stochastic
Alert sent to
Meta-brain-BB
individuals
involved
PRINCIPE DU MONTAGE
GNA
VOLTS
1
Secondes
0
0
150
Terrorist Attacks, Sept 11, 2001
electronic component resonance
World trade Center Attack

The chronovisor

A device called the "chronovisor" was developed there is half a century by a team of scientists—including Fermi (atomic physics specialist) and Wernher Von Braun (father of the US space program) became interested in it—gathered around a Benedictine monk of Venice, Padre Ernetti, specialized in Gregorian chants, and graduates of quantum physics.

This machine would have worked in 1972, capturing the wave bands and managing to "visualize" holographic scenes from the past:

Napoleon's speeches, Cicero's "Quousque tandem Catilina", climb to the Golgotha... All this was filmed, presented by Pope Pius XII and the highest Italian civil authorities.

Father François Brune, who knew Padre Ernetti, relates that the holographic characters were not very large. Roughly the size of our television screens. You could set the device on the place and time that they wanted. They chose someone they wanted to follow.

It is on him that they regulated the device and then it automatically followed. The chronovisor was described as a large device with branches of unknown metal alloys, connected to a cathode ray tube.. *(SEE ARTICLE MORPHÉUS. FR JOURNAL March 2012)*

The images produced by the "chronovisor" are not copy-paste of existing images. This is also what has been shown by analyzing digital photos of holographic projections by Nancy Talbott of the BLT Research.

Thus the sculpture of Jesus Christ from the Sanctuary of Collevalenza, Italy, done in Spain in 1931 by the sculptor Lorenzo Valera Cullot is stored in the planetary-BB's data. The image projected by the "chronovisor" is true, but no identical to the original sculpture.

The assumption is that the antennas or the machine itself contained elements based on germanium and silicon.

The Padres Ernetti Gemelli and are connected to the planetary-BB, like everyone else, and pick up subconscious information. By focusing on the mental image "of a crucified figure" they exert a selective filter of the corresponding mental images in the planetary-BB. Their biofrequencies are transmitted and amplified in "chronovisor".

When the targeted person is identified, the operator locks the frequency of "chronovisor". Only mental images corresponding to this frequency are then holographically projected by the device.

The device projects images resulting from planetary-BB, but transformed by the operators mental projections according to their ability to produce them. If other people had done the research, "Jesus Christ" with the "chronovisor," it is likely that the result-ing projections had been different.

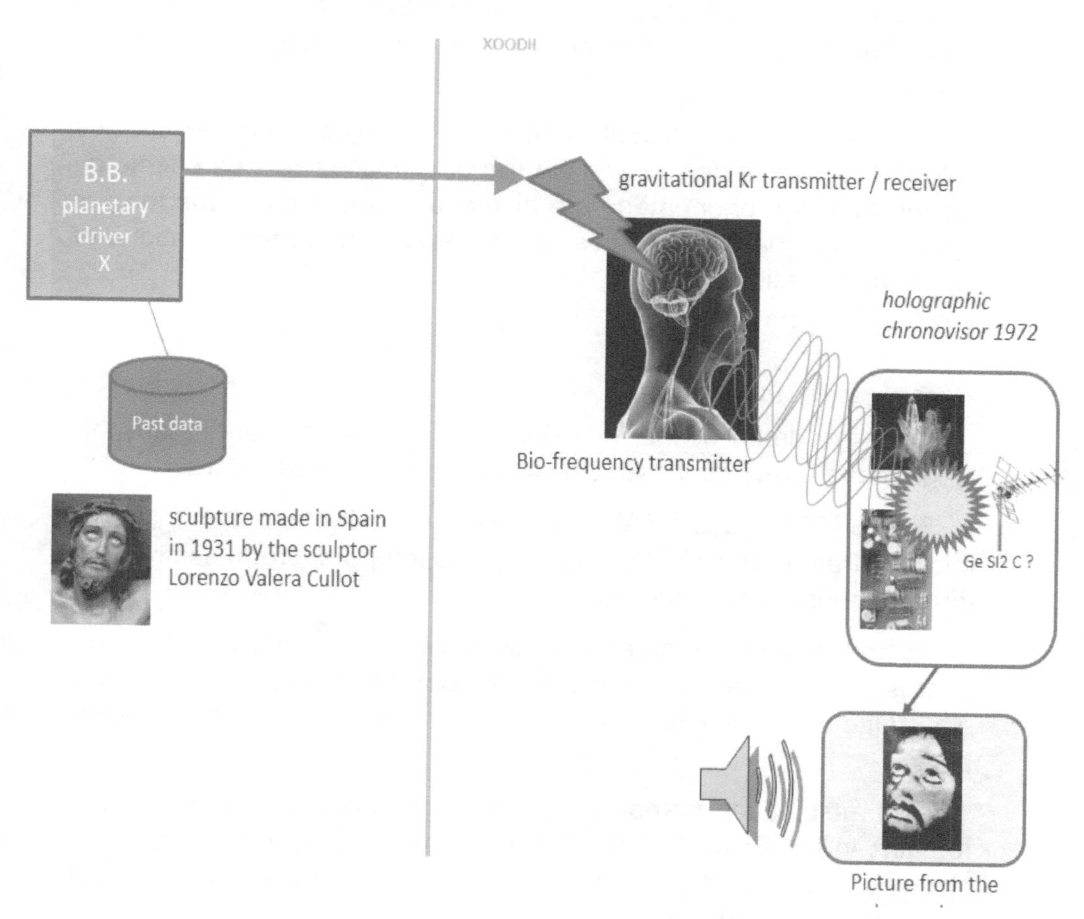

"HAUNTED" PLACES

We have seen six type cases of the explanation model with the cosmological context. All possible combinations of these combined configurations can produce similar effects with different causes.

We presented the effect models whose causes have in common that they all require one or more receivers who transforms and project the received stream.

All these projections phenomena are expressed in the same term of "ghosts or paranormal activities". There are also, all the cases events where the cause of the phenomenon is directly due to the residual Aura.

The deceased's residual Aura directly projects the frequencies onto the electronic device, which talks or produces various effects on matter. All these phenomena are also expressed in the same term of "ghosts". The "haunted" places are generally direct manifestations of one or more residual Aura.

SPACE-TIME TRAIL AND AURA

With the death of the individual, the spatiotemporal-human, this long "snake" loses its "head" which is the physical body of man.

The SPACE-TIME TRAIL engraved by OEMMII still exists in cosmic substratum IBODSOO, but it stops its temporal progression in the three-dimensional framework.

In his lifetime the man emits multiple bio-frequencies as aura. This aura is also inseparable from spatiotemporal-human that the physical body. Under normal circumstances, aura disappears with the physical body.

The assumption is that under certain conditions, a residual aura remains in the SPACE-TIME TRAIL. This aura is a residual electron cloud, probably cold plasma. This residual aura could evolve in different ways, depending on the case.

The two main scenarios:

- the residual aura is absorbed when the Soul is integrated into the planetary-BB
- the residual aura stays in the SPACE-TIME TRAIL

The spatio-temporal Man linked to his aura bio-frequencies and BUAWA

INCIDENTS OF INTEGRATION OF SOUL-BUAWA TO THE PLANETARY-BB

Normally, the body-OEMII/soul-BUAWA link is broken with the death of the individual. A LEIYO effect then connects soul-BUAWA to planetary-BB on which it depends.

The assumption is that this effect LEIYO absorbs some human energy fields, bio-frequencies of the aura of the human body before breaking the body-OEMII/soul-BUAWA link.

In other words, in a normal case, part of the energy of the aura of the deceased, helps to initiate the process of integration to planetary-BB.

According to some "mediums" this would be an average of 4 Earth days to be partially absorbed by the integration LEIYO effect and the rest would be scattered like any electromagnetic field.

When a person dies in great psychological distress, his aura emits a high level of abnormally high bio-frequencies.

This aura, abnormally energetic, permeates a certain level of energy in the 10D-OAWOO frame, on the last segment of the space-time trail drawn by the deceased. The energy is used for the stress signals and there is not enough energy for the integration process. So, the aura of the deceased will remain blocked in the space-time trail.

Sometimes the "ghost" has its little habits and rambles a bit, he is seen repeatedly in the same location. It's always playing the same "film" projected into the same room, even if others "films" can be played in this room, sometimes overlapping each other...

This is actually what it should be. A "film" was produced by humans, often in distress, and they strongly expressed bio-frequencies in the terrestrial environment, the last segment of the space-time trail engraved before death and residual aura remained trapped there.

The assumption is that under certain conditions, the trapped-in residual aura into the space-time trail may be perceived by a human being.

Some people have the ability to "see" the "ghosts" in cemeteries, in premises where a historical drama is known or even to places that seem arbitrary...

Some people "mediums" have the ability to "see" the information from this part of the space-time trail. We believe that these "mediums" have two capabilities:

- one to issue bio-frequencies
- the other to see or feel the bio-frequencies

These people have a close enough ability of the radar principle. A wave is emitted by the radar, it bounces off the plane and it is returned to the radar with a slight deformation.

In our case, the wave bounces off and reverbs the residual aura and is returned to the person who then "sees" or feel the echo of the residual aura.

According to the energy levels of this echo the residual aura will shape a pseudo-hologram following the 10 axes of IBOSDOO. The echo can also return frequencies in the field of visible wavelengths or in a field of recordable frequencies by a digital camera sensor, in the high ultraviolet or infrared, for example.

It is possible then to photograph or see "the ghost hologram" of the residual aura which was stored in the past in the space-time trail.

The residual aura

The residual aura of the deceased is blocked in the space-time trail and secondly the integration process of BUAWA is also stopped. The residual aura remains bound to the BUAWA. Although the integration of the Soul is blocked, it is likely that residual aura also remains bound to the residual BB-planetary, but without integration processing.

The hypothesis is that the LEYIO effect of BUAWA integration to planetary-BB does not occur correctly. Instead of connecting to the planetary-BB, the connection is made with the electronic mist produced by the residual aura, highly energized by the body during the traumatic incident.

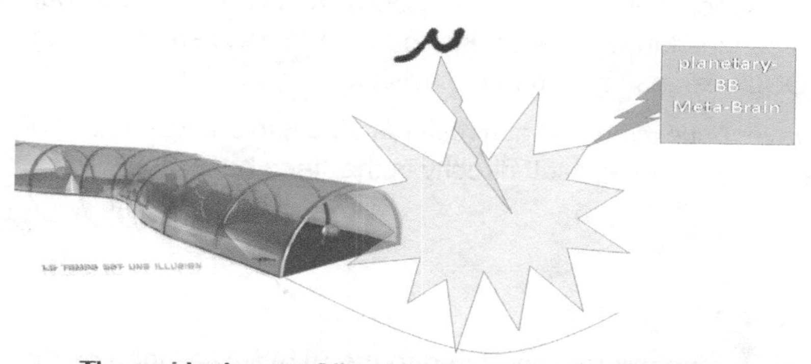

The residual aura of the deceased remains trapped in the spatio-temporal framework. The BUAWA was not integrated into the planetary-BB and remains connected to the residual aura and controls it.

BB link with the residual aura

The integration process of BUAWA was stopped. The residual aura remains bound to the BUAWA. Although the integration of the Soul is blocked, it is possible that the residual aura is also bound to the planetary-BB. However, this link with the planetary-BB is blocked or frozen...

This would explain that the integration process of the BUAWA can be revived later.

The antenna effect with the XOODII cosmos layer

As we have seen in the experience of the Global Consciousness Project, the electronic fog is an issue of frequencies covering the entire electromagnetic spectrum and which creates an antenna effect with the XOODII cosmos layer.

It is this type of device which was probably used for the chronovisor.

Another mounting can also be set up to transcommunication receptions experiments, using an old analogical television without antennae and connected to a camera.

The television produces an electronic fog—the snow on the screen—and the camera films this fog and feeds back it into the TV. This "feed back loop" will take place at "infinity" and capture the global-BB flows (in connection with the integrated BUAWA) transmitted them, via the XOODII cosmos layer to the television.

It is however not excluded that a residual aura can come to interfere by producing a resonance, itself directly in the "feed back loop" ...

The ghost's life

The residual aura will remain bound to the BUAWA, as long as the soul-BUAWA will not receive the energy necessary for its integration into his planetary-BB...

The cold plasma, what probably the residual aura must be, need to maintain its energy level by absorption of assimilable energy by it: the human bio-frequencies. This is the active "ghost" of the deceased.

We could say that the pair residual aura/BUAWA is a kind of OEMMII AIOOYA Ammie, a human disembodied. If he has enough energy this "ghost" may continue to evolve outside the space-time trail engraved by the deceased OEMMII.

The pair residual aura/BUAWA blocked in space-time can enjoy an antenna—a OEMMII medium—to manifest itself in various projections. It may make projections mass, holographics volumes, or both in extreme cases...

Projections in "saving energy mode" will be one way to give the ORBs (*see subchapter ORBS*).

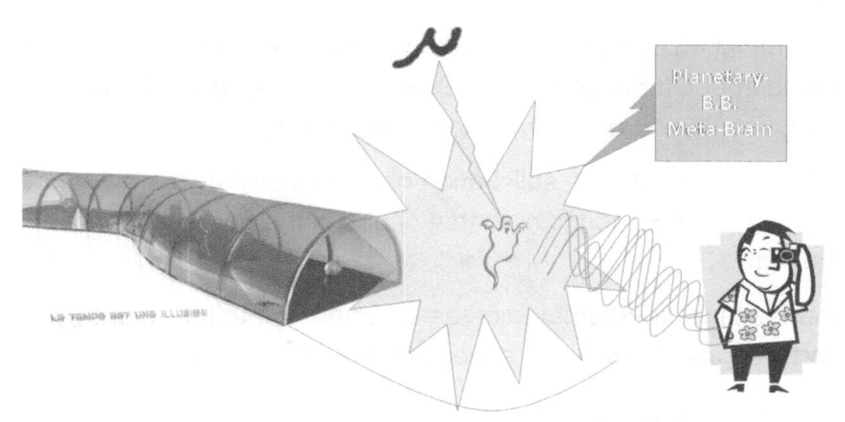

**The residual aura bound to the Soul-BUAWA increases is energetic level by the absorption of human bio-frequencies.
This is the active "ghost" of the deceased.**

Presentation of a case of residual aura

la Penne sur Huveaune — France — photo © Michel Marcel

This photo apparently harmless, is not. The flash reflects a task that is visible only on digital photography. A simple improving contrast and brightness show that there is a problem with the wall behind the table.

The examination of the wall surface is perfectly smooth, the grazing light gives no effect, measures a posteriori in the infrared and ultraviolet not give any results, no electromagnetic field is detected.

The story began with the solicitation of a "medium" by the owner of the place. It was concerned to note that chairs of the house had moved during the night...

On the spot the "medium" indicates is connected to the "spirit" of a person tortured and thrown into a well at the end of the Second World War.

The well will be identified under a slab of the basement of the house. A tape recorder is switched on in a room of the house, and then the "medium" does his psychic work to disconnect the residual aura in the cellar.

The tape registration will deliver a message of thanks to "medium". The electronics and physical engineer-researcher, Marcel Michel will be pushed unconsciously without any rational reason to go into the salon owner take the picture presented here against.

Interpretation of the case

This beautiful classic case of school, treated in a very professional way is widely generalizable to many similar events, and it can be explained by the proposed theory.

It seems that the residual aura has enough energy to move and make demassification projections onto the chairs to signal its presence to the owner. Who reacted well and wisely seek the help of a "medium".

The intervention of the "medium" brings enough energy to boost the integration process of the soul-BUAWA to the planetary-BB. This results in disconnecting the residual aura of the impregnation zone which moves to the salon before the integration process.

The measures are carried out after the residual aura integration or after it moves out of the salon. This may explain that the measures give no result.

ORBS AND ANIMAL PROJECTIONS

Sometimes ORBS seem to have animal figures or that are reported testimonials of holographic projections from pets.

As shown in the Oommo documents, animals have no connection to the Soul-BUAWA. Yet in an indirect way, they integrate the Meta Brain.

The planetary BUUAWE BIAEEI contains information and perceptions of mental processes of OEMMII superior beings. A human attached to a pet will get into his soul-BUAWA and into the Meta Brain, a huge amount of information characterizing the pet, its image, its sounds, the emotional perceptions, etc.

When the soul of the deceased human integrates the Meta Brain, it will find all the information in this environment including the pet's information...

The animal's intelligence is purely neuro-cortical, which goes to the australopithecine primates, therefore already highly developed... In the dog's case, some perception of its environmental information can be found in the planetary Meta-Brain via the intracellular krypton emitter. It's the animal itself which has rebounded flows through its environmental perceptions to the emitter-receiver BAYODUU-cell. This flow is supplemented by the information conveyed by the master of the animal. The aggregation of personal information of the integrated soul-BUAWA of the deceased master, combined with multiple additional information, make the picture of the deceased dog "visible" by his deceased master.

The integrated soul-BUAWA of the deceased master continues to accumulate information forever. While dog-related data will be complete and finished in the planetary-BB.

THE "POSSESSIONS"

Sometimes during OUIJA sessions in particular, the flow from a soul-BUAWA remains connected to the receiving person. Popular traditions speak of "possession". This type of "possession" is "Transcommunication" type.

The same phenomenon of "possession" may result from the action of a residual aura. Probably disturbed by a living human, it will "wrap" aura of living human, and thus take control of the living body.

To break the flow still connected to the receiving person or eliminate the residual aura "possession" requires an energy input.

That is what that seems to know some people ordinarily known as "exorcists". This is a "violent" contribution of bio-energy frequencies. Other people rather termed "mediums" operate by "soft" contributions of bio-energy frequencies via incantations or "prayers".

In the case of the residual aura, the energy of the "medium" will reconnect the BUAWA to the planetary-BB. If any, the rest of the residual aura will disappear like any electromagnetic field by lost energy.

THE POLTERGEIST

The poltergeist is mainly an effect on the masses that could be generated by a residual aura.

As we have seen for the effects of instrumental transcommunication, it operates by resonances and electromagnetic induction.

This between electronic devices and inductive electromagnetic devices or resonances with frequencies emitted by the human brain activity or even by the emission spectrum of a residual aura.

This residual aura could therefore generate this kind of resonance and electromagnetic induction, and perhaps also a micro LEIYO fact that generated a relative "demassification", that is to say a gravitational resonance effect on the targeted object due to very high potential differences.

The residual aura produces a range of frequencies, more or less pulsed, which resonate with an object or simply produce an electromagnetic or electrostatic discharge. The move thereby can also produce a noise or just a local deformation of a rigid structure. This is torsional effects of the axes of the masses, as we have already presented.

Exocivilisation's residual aura in Easter Island

During a trip to Easter Island in 2014, the magnetizer Gilbert Attard and the "medium" Monique AUBERGIER face atypical phenomena they describe on the website of the Association CROPS.

Facing some MOAI, they are subjected to powerful "telepathic" flows. For these flows the MOAI transmit them, information on an ancient culture of the Easter Islanders, and they themselves say, we MOAI, are "alive" and we communicate.

The information flow was made in French telepathically, and "channeled" in Pascuan language during the presence of the Easter Islander guide. Gilbert Attard directly vocalizes the mental stream received in that language he doesn´t know...

My hypothesis is that each MOAI is "surrounded" or "contains" residual-aura. But they are not human's residual-auras.

In fact, it would be of residual-auras of an exocivilization. These, very energetic, could directly connect on human auras and thus reach the mental connection centers.

It is a connection of an exocivilization's residual aura to humans Gilbert Attard and Monique AUBERGIER. It is likely that the flow can also be achieved via a telepathic channel as in the case of "channeling".

These would be an old exocivilization's residual aura, previously present and described in the tradition of Easter Island and on petroglyphs of "giant's long skull".

Possibly, some of those people decided to die without integrating their Meta-Brain and leaving their residual aura keep the place.

This assumption is reinforced by the experience of the cave Vai Teka where Gilbert Attard and Monique AUBERGIER detected some exocivilization's residual aura that also protected this place.

The residual aura of the exocivilization of "giant's long skull" seem to have very different capacities then the terrestrial ones. Indeed, a particular MOAI with eyes, projects a significant flow onto Monique

AUBERGIER. This has the effect of increasing its sharpness, probably in the high-UV frequencies and allows Monique AUBERGIER to visualize alien aircraft around. She sees multiple "flying saucers" that other tourists in the group didn't see...

These machines are probably encapsulated by a photonic bypass field in the frequencies of the human visible, to make them invisible to terrestrial humans. Things partially managed by the Earth science.

By increasing the sharpness of Monique AUBERGIER the MOAI with eyes wanted to give him the opportunity to see these devices.

In this case, three distinct exocivilizations are currently present on Easter Island.

Wavelengths of "ghosts"

Without making an exhaustive history, of technical and research on bio-frequencies related to the human aura, nevertheless we can give a little insight.

In 1936 the researches of Roy Davis and Walter C. Rawls lead them to identify the human biomagnetism. Numerous studies and developments in this area will address bio-frequencies, but there are others. In 1939 the Soviet Semyon Kirlian and his wife Valentina take photographic images that show a halo of light around an object subjected to high voltage. This luminous halo is explained by an effect called "corona" of gas ionization, caused by the immediate vicinity of the object immersed in a strong DC or AC electric field. Plasma is then created and the electric charges are propagated passing ions to neutral gas molecules, the fluid becomes ionized and becomes conductive plasma. This effect also depends on the diameter of the object, its surface condition, its density and humidity of the surrounding air. It is not excluded that the Schumann resonances of 3–30 Hz electromagnetic field of the Earth also have an impact on the effect corona.

In 1983, Indian researchers Kejariwal, Chattopadhya Choudhury and showed that the effect "corona" easily occur with frequencies above 100 kHz and a voltage between 15–20 kV.

Many devices for measuring electric fields in human frequencies were invented. The normal frequency of brain electrical activity is between 0 and 100 Hz and the maxima in the first third of the strip. That of the muscles frequently rises to around 225 Hz, and the action taken on the heart gives about 250 Hz, but this is the ceiling for this electrical activity associated with biological functions. Note that in the electric fish frequencies are also in the order of 250 Hz.

As we mentioned above, Valerie Hunt used an electromyograph which is a diagnostic technique using electric currents. The bio-frequencies of the Aura went from 100 to 1600 Hz. In addition, instead of emanating from the brain, heart or muscle, the electrical activity intensified over areas traditionally associated with the chakras. She also discovered that each color in the aura of a person fit a frequency curve that she learned to associate with an oscilloscope and it gave a translation on the screen. Furthermore, eight "mediums" could simultaneously compare their perceptions with those of their colleagues and the graph generated by the oscilloscope.

The broadcast and transmission of information via human bio-frequencies is also an established fact in multiple wavelengths. In 1962, Professor Anna Gurwitsch, through the use of photomultipliers confirms the existence of biophotons. Work continues with Terence Quickenden, Shane What Hee in 1974. Fritz-Albert Popp, coined the term, defines the biophotons by the intensity of their emissions on the surface of living tissue, which is in the range of 10–1000 photons per square centimeter per second.

The typical magnitude of biophotons is in the visible and ultraviolet spectra, and biophotons are involved in intracellular metabolism.

It is possible that the intracellular information flows, chemical, magnetic, electrical, scalar, photonic are more or less redundant for safe functioning of all sophisticated metabolism, requiring numerous coordinated interactions... These resilient mechanisms are usually used in industrial technologies of control and steering...

The holograms of residual auras can be captured by digital cameras. In addition, generally the energy produced by the residual auras inhibits cadmium sulfide photoresistors of the flashes.

This means that these "holograms" return frequencies in the ultraviolet or infrared. Children see residual auras more easily than adults. The spectral efficiency curve of children of 2 months compared to adults, shows that the children have a higher sensitivity in short wavelengths (400 to 500 nm) of 0.3 log units, than in adults (DOBSON 1976).

It seems that dogs and cats see in the ultraviolet, also perceive the energy produced by the residual auras.

Infrared spectra and ultraviolet of photoresistors of digital cameras

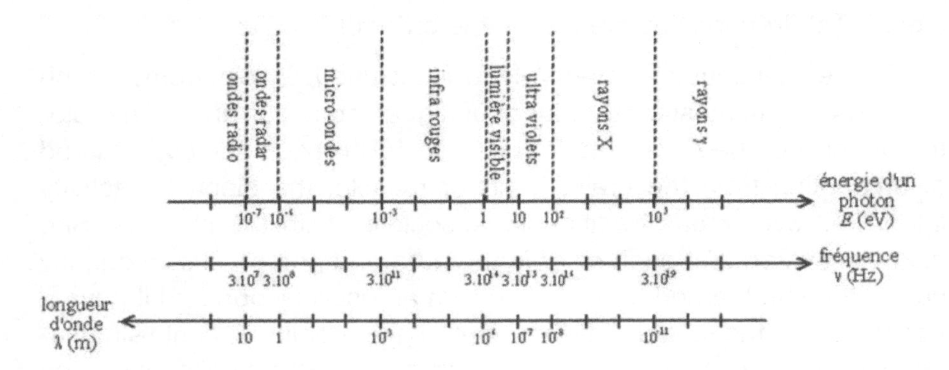

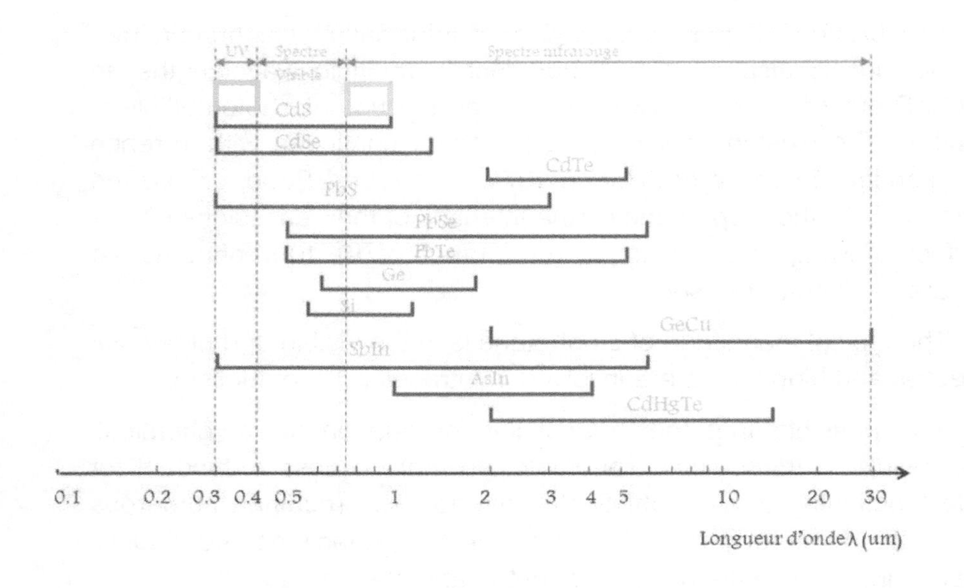

Scalar bio-frequencies?

We see with these multiple systems as the human auras, and the residual aura, are at least partially detectable and measurable.

We suspect that these residual auras also produce bio-scalar frequencies, the effects of which are called LEIYO by our friends from UMMO. These scalar bio-frequencies allow the "switch" or flip of dimensional axes to cause various phenomena, such as objects demassification, usually called psychokinesis, etc.

We previously mentioned Gilbert Attard and Monique AUBERGIER trip to Easter Island, the tradition says that MOAI moved from the quarry of the volcano to their place of installation, and standing alone.

My idea is that the Easter Islanders used ritual songs as frequency carriers to issue scalar bio-frequencies. What they call the "Mana" that individually some islanders have.

By the fact of psalms rhythms, scalar bio-frequencies come into resonance and produce switches onto the OAWOO axes of mass of the MOAI, allowing them to move in small leaps. MOAI weigh some 250 tons...

It should be noted that the "Mana" is also a capability known and used in Africa.

The "ghost" in minerals

Storing information on the spatiotemporal trail is potentially eternal. A residual aura can remain trapped long. The information transmitted by the bio-spectrum of a person, regardless of their condition, may also potentially be stored in the crystal structures of minerals. The storage is very stable, it can be stored over very long periods.

Holographic flow diagram: storage and playback

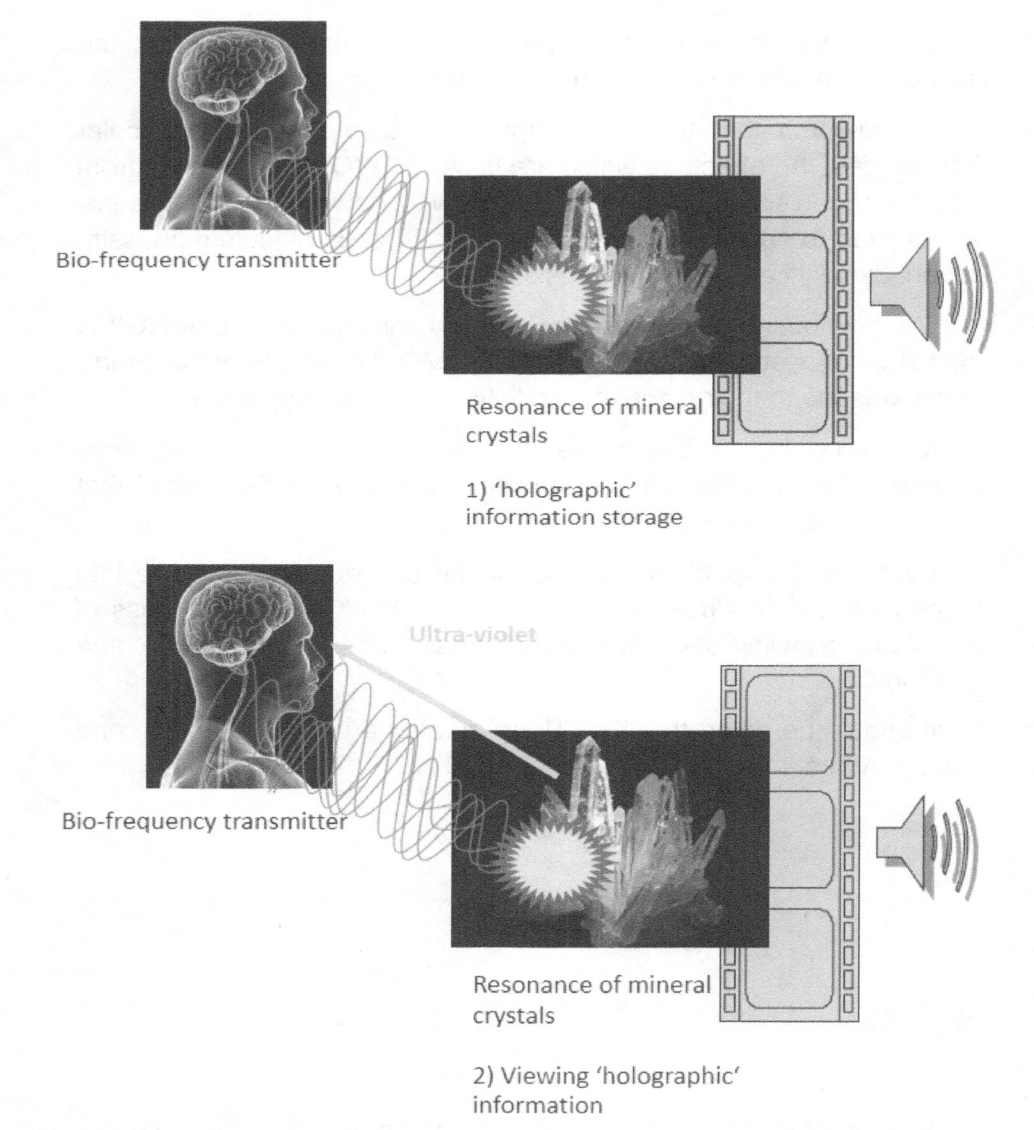

The orbs

The arrival of digital sensors in cameras was accompanied by light reverberation phenomena on dust and various chromatic aberrations. However, these causes do not explain the nature of all observations.

For example, in a closed dust-free place, photos taken in an empty room, where there is no ORBs.

When in the room there are people who realize sustained mental activities, it can manifest "bubbles" on clichés even without flash.

The hypothesis is that these kinds of ORBs are projections which are formed according to the different types of projections presented above.

Some result from bio-frequency projections via a relay-individual and others are simple holographic projections derived from a mineral storage...

But these projections have in common to have a low energy level. This limits the phenomenon to reduced manifestations. When using flash, the reverberant energy permeates the photo sensor and reveals the shape.

We can find either projection giving low resolution faces or just opaque "bubbles". But even in this mode "saving Energy" Transcommunications recordings are reported and some "mediums" report being in connection with the ORB.

Example of several projections—that of the center is enlarged—September 2012 in the Chapel St Jaume in Carcès—France—photo © Marcel Mylène

Crop circle and "ghosts"

My friends Nancy Talbott and Chris Cooper contacted me to collaborate on cases of Crop Circle, following a case of photographs of various "holograms" with various characters, including those of a blue-indigo E.T. and a spacecraft.

These "holograms" were they done by our friends from outer space?

For example, the makers of crop circles that we had identified in Presence, UFOs, Crop Circles and Exocivilizations?

Was there a link with the crystals of pure hydro magnesite dolomite found in certain formations?

So I made the assumption that the "holograms" resulted from storing information in hydromagnesite crystal. However, one place was in this case.

www.robbertvandenbroeke.com and www.bltresearch.com

All pictures shown here are taken with the camera of Nancy Talbott near by the "medium" Robbert van den Broeke.

Then we made the assumption that the "holograms" were related to Crop Circles themselves. But many photos are well outside the formations.

I asked our team of experts to analyze and compare the image of blue indigo E.T. and the makers of crop circles that we had identified in Presence, UFOs, Crop Circles and Exocivilizations. But these are not the same characters.

I conclude that it is the presence of the "medium" Robbert van den Broeke is the cause of these projections of "holograms". There is therefore no link with the Crop Circle themselves, nor our friends from outer space.

Near dead experience—NDE

NDE or Near Dead Experience are the living proofs of the proposed cosmological model.

In this state the brain bio-frequencies are stopped, but the complex of krypton atoms of OEMBUAWA continues to receive enough energy to continue to function and begins a LEIYO flow for the integration of soul-BUAWA to the—planetary-BB.

We will see an explanation of a specific case that two distinct phenomena may occur.

In 1991, in Phoenix, Arizona, the Pamela Reynolds case is fully documented in a surgical procedure to give her clinically dead for over an hour. Dr. Spetzler sets up a "hypothermic circulatory arrest" at 15.5 °C and drains the blood to treat the aneurysm. Under these conditions the body can survive normally between 30 and 60 minutes.

Pamela Reynolds recounts being out of his body and she had observed the whole scene from above. The surgeon Robert Spetzler attests: at this time of Pamela Reynolds operation was clinically dead ... but the complex of krypton atoms of OEMBUAWA continues to receive enough energy to keep running.

Initially, the aura of Pamela Reynolds detaches from his body. His aura remains connected to his soul-BUAWA and BB-planetary. The aura captures a set of frequencies including bio-frequencies of medical and technical persons in the surgery platform. The aura sees, hears and transmits the data to his soul-BUAWA and planetary-BB.

Secondly, the aura begins the integration of the soul-BUAWA of Pamela Reynolds into the planetary-BB on which it depends. The soul-BUAWA can begin to exchange information with other souls already integrated into the planetary-BB.

The BB-planetary decides he wants to integrate information. When there is no more bio-frequency brain but all the complex krypton atoms of OEMBUAWA is still operational, it is possible that the planetary-BB denies the connection to a soul-BUAWA and tries to reconnect it to his body.

If the reconnection succeeds because the body has recovered enough energy, and that the brain's neurological structures are operational, the subject will resume his earthly consciousness. Otherwise, unfortunately the subject will remain in a vegetative state with an operational OEMBUAWA, but not operational brain structures.

The residual aura-memorial cyst

Sometime after the death of his father, my friend Frédéric began to suffer from the left hip. So much so, that soon it moves only with difficulty with a cane and a corset. He did not 40 years and thorough medical examinations are formal: it has absolutely nothing to hip and is in perfect health...

Is it a psychosomatic pain due to his father's death?

The pain is real and handicapping. Philippe Douillet, our friend also expert in geometry, also has abilities to perceive auras and residual aura.

It finds that a residual aura-memorial cyst is connected to Frédéric's left hip. It also emanates information from this residual aura-memorial cyst.

This would be the manifestation of a deceased person during the Crimean War to 1853–1854. The deceased was a soldier of the expeditionary force sent by Louis Napoleon Bonaparte, who received a piece of one of the first explosive shells, fired on this occasion. A burst would have hurt the victim of the left hip and caused his death.

The residual aura was in the form of a memorial cyst emanating from the back of Frédéric's left hip. Philippe described as resembling a mold fungus, fungal from 70 to 80 cm long with at its end a "head", huge as 2 fists...

Although immaterial, Philip said that the head looks like a compact tangle of small dark cords (color between bitumen and very dark chocolate), as hard as a ball of string, but consists of many pieces. Each piece seems to him a piece of memory, a poorly lived event. In Vedic culture, this is also called it a karmic residue.

My hypothesis is that the violent death of the soldier halted the integration of the Soul at BB. But a particular emotional context, the parameters have not been established, have structured the residual aura as a memorial cyst. The latter being so bound to the soul of the deceased, and the connection to BB stopped in its integration process.

Although we do not have accurate information about this, we can imagine that the residual aura of the soldier could have remained on the battlefield, and by chance and opportunity, she would have found a carrier or conveyor... Perhaps it was a comrade who survived and later returned in contact with the family of the deceased. In doing so, the residual aura be transported and be able then to connect to a living being familiar...

And passing through generations of a family member to another, this residual aura-memorial cyst would be attached to Frédéric's father. Upon the death of the latter, will the residual-would have changed "carrier".

Frédéric later confirm that one of his great-great uncles died during the Crimean War...

Ultimately, Philippe Douillet was able to transmit enough energy to this residual aura-memorial cyst to detach it from Frédéric and be disintegrated in a normal reconnection process of the Soul of the deceased.

Philippe uses Tibetan bowls whose harmonics will serve as carrier waves for its ability to project bio-scalar waves onto the residual aura-memorial cyst, whose trauma is recognized and "let go" by absorbing the energy to restart its final integration process at BB...

"When I had intuitively found the right bowl, good sound of this bowl, the right position, the right angle, the memorial ... cyst began to unravel. To become less compact, less opaque. The various pieces of "cord" have begun to unravel. Both take more space and become less gloomy. The phenomenon continued, the "strings" parted increasingly from each other, taking more volume ... but without change of position relative to each other.

And they also became increasingly bright.

Initially the memorial cyst was as big as two fists, then like a balloon. He moves from handball size to basketball size and bigger... The "strings" became not only bright, but more and more transparent. In the end, all becomes so voluminous and transparent as it disappears to my "view".

Conclusion

Communication with the "spirits" portrayed as a subject "paranormal" or metaphysical, is rationalized with the cosmological framework described in UMMO documents.

We believe that the amygdala, the pineal gland and subtalamiques nuclei are the key brain areas for connections to soul-BUAWA and BB-planetary, and for of human bio-frequencies.

There would be three main classes of phenomena:

• the connections of integrated Soul-BUAWA of defuncts in planetary-BB that are at the origin of human projections of resonant bio-frequencies, are "transcommunication" type phenomena.

• the residual-aura phenomena also produce resonances effects giving the main "ghosts" and poltergeists

• the holographic storage in the minerals that is visible by human bio-frequency radar effects, produces certain ORBS and "ghosts"

The projections are the result of LEIYO effects of twisting dimensional axes AOWOO. Those are energy projections of masses, or holograms on volumes, or both.

Although their causes are very different, most of these phenomena are expressed in the same terms of "ghosts".

The few standard models that I imagined by prospective allowed me to make a streamlined way to all cases which were submitted to me so far.

Some cases are really extraordinary, it takes a lot of patience and kindness to keep an open mind both critically and constructively to examine these cases. A conceptual framework through streamlined cosmology allows this analytical approach.

Ultimately, it is not God who created man in his image, but man that fills the planetary Meta Brain of his images.

We can hope that in the coming decades we will have the technical means of experimentation of these theses in a public and civil framework...

*

MENTAL HANDLING AND ABDUCTIONS

God and Exocivilizations

In "Presence 1—UFOs, Crop Circle and Exocivilizations" in 2007, we presented a detailed summary table describing 18 exocivilizations.

Since, additional information, also confirmed by our Oommo friends, has enriched our knowledge about it.

In 2015, so we have 23 exocivilizations installed on Earth or on the moon. As we explained in the book Presence 1, visits have different scientific objectives, as part of a "Pax Galactica" that is at work.

This is ensured by (at least) an exocivilization, very old and very advanced (significantly more than 23 exocivilizations who visit us) and it guarantees our general cosmic security. In other words, no exocivilization can invade Earth or exercise warlike activities against the entire planet.

As we have seen in the previous books and following our friends of UMMO, among these 23 exo-visitors have experimental objectives very varied, and their impact on the terrestrials too.

Even if their behaviors are in the ethical limits permitted under the "Pax Galactica", perceptions of these limits depend on the psychology of our visitors.

The behavior of our visitors will also depend in part on their physiology. For example, some of which may have a physiology that absorbs human bio-frequency, making them potentially harmful if they have a physical contact with us. We could be "discharged" as vulgar batteries...

But potentially the most adverse impacts are more to fear from a psychological having a lack of control or recognition of all objects of universal cosmology.

Indeed, all exocivilizations who master the cosmological panorama we presented, necessarily including interdependencies between exocivilizations. This is summed up by: all that is bad for us, is bad for others exocivilizations.

Nevertheless, among our 23 races of visitors from outer space, 4 of them do not recognize all objects of this cosmology. Let's look at the implications:

As we have seen in detail, knowledge of multi-cosmos WAAM-WAAM structure is essential to master interstellar travel. But that has no obvious impact on the psyche of our visitors.

About WOA, "GOD" among the infinite number of waves that it transmits in the Universe through the planetary Meta-Brain, diffuses ethical laws. Not recognizing this transcendent entity, probably lessens their impact, but not the overall perception of ethics that underpin them.

Our visitors thus have their specific understanding of the general ethical frame. It is certain that the 19 benevolent exocivilizations that are on our soil adhere to this concept. The majority of these seem to practice Metaphysics, or ritualized religion around the concept of God, this probably related to their cultural history. This is the logical evolution for humans ... in all the cosmos.

Knowing AIIODII, the "Absolute Reality", the XOODII WAAM, the "inter-cosmic layer" is probably implied for travelers of the cosmos.

For example, without this different measurement systems for their trips could not be developed at an advanced level.

Knowledge of the WAAM-UU, the "Meta Brain Cosmic" and "global brain" BUUAWE BIAEEI is a sensitive point.

It allows to take into account the positive or negative interactions of a civilization onto another, at a cosmic level. Do not master this aspect might suggest an exocivilization, that its actions are independent of the others, and so that her experiments on others does not impact.

We perceive here the risk of slippage... A misunderstanding of the WAAM-U, the cosmos of "Ames" and BUAWA, "Soul" obviously have the same risks...

It is one of those cases that concern 4 exocivilizations on our soil. 2 of them we can synthesize more accurate information.

We first see the case of GOHOians involved in a scientific program including abductions of terrestrials, under secret agreements, illegal and illegitimate, with the US military.

In Chapter 12, "*The Anunnaki and the Reptilians*" then we will discuss the likely role of the TWO-ians, often mentioned in the welter of rumors and misinformation to as "Reptilians".

The abductions

We reported in "*PRESENCE 1—UFO AND CROP CIRCLE EXOCIVILIZATIONS*" information relating to abductions.

The exocivilization of GOHOians realized since 1948 a study of physiology and psychology on terrestrial humans, largely on the territory of the USA. It seems that few other exocivilizations have punctually similar practices.

The physiological study of terrestrial humans is done without major injury on the subjects. This dual testing of physiology and psychology on long-term has not gone unnoticed by the consequences of the psychological study of people.

It is told through many testimonies relating to exocivilizations "SMALL guys with LARGE HEAD AND BIG EYES".

Basing us on the data presented in "PRESENCE 1—UFO AND CROP CIRCLE EXOCIVILISATIONS", we can roughly estimate that 70% of individuals are exocivilizations "little guys with big heads," and one of them is probably the species described by Budd Hopkins.

Portrait of a robot GOHOian by G. Cousin

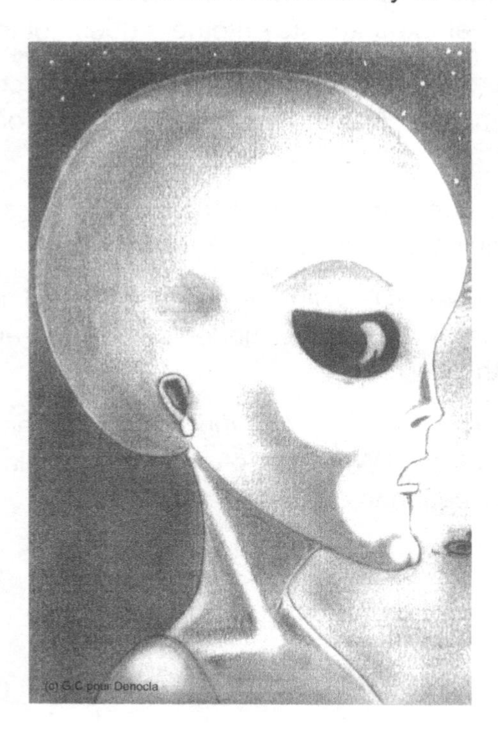

About a thousand cases a significant number of subjects relate to having seen a UFO observation.

Mental manipulation and memorials flashing

The hypnotherapist Yvonne Smith shows that subjects have standardized memories.

While the UFO is generally described as being a few tens of meters in diameter, the subject describes a situation where it would have been led to huge laboratories inside the spacecraft. Other recurring themes are those of “medical analysis” and “the manufacture of hybrid beings.”

In a theoretical absolute compression volume is possible for intercosmic travel, under very specific conditions, but the “abductees” are in our conventional terrestrial dimensional framework with the classical laws of physics.

The inconsistency of the contents of memory with standardized dimensions actually observed in the size of the craft can let us think, that some of the psychic experiment could be the transfer of false memories into their brain.

We may think that some of these experiences are mainly to test critical analysis skills of the subject facing situations posing ethical questions...

This implantation of false memories standardized, this "memorial flashing" is deleted from the conscious memory of the subject after the experiment. Bad luck for the experimenters from outer space, hypnosis techniques have revealed the abduction and the mental manipulation.

In specific cases, the "memorial flashing" transfers knowledge or new skills to the subject: the ability to produce atypical electronic achievements while the person has never had any knowledge on this subject, the ability to produce drawings very developed whereas before the person has never had this skill...

The real abductions are probably fewer than the many testimonies may suggest, but the "memorial flashing" seems to be very frequently used by various exocivilizations. Examples.

THE "MEMORIAL FLASHING" OF KNOWLEDGE

In 1996, Soissons in France, a van of the police makes a night round in the campaign. The van stopped itself on the road, can neither advance nor retreat, near a strong light behind a bush. All police say staying conscious but "paralyzed."

One of them, Jean-François M. gets out of the vehicle and goes behind the bush. He will remember nothing, but colleagues said he left at least 30 minutes. From that moment his life changed, he was attracted by technologies that knew absolutely not—electronic, engines—and he seems to have spontaneously acquired knowledge of these techniques, without any training of any kind...

He realized electronic assemblies virtually taking of components "random" without knowing anything about it, not knowing himself why he was operating well. But it worked! This neophyte in Science and Technology will file several patents with the INPI (patent deposit).

He perfected the exceptional water engine of Pantone, getting 87% water for 13% gasoline!

Patent No. 0902947 repurchased and swiftly buried by a car manufacturer, which will fall into the public domain in 2020 ... maybe.

The "memorial flashing" of ability

La Ciotat in France, 1976 Claude G. is driving near the Montagne de Lure in the evening. At the top of a hill, a very bright ball comes near his vehicle, which swerved it and affects the driver by sharp tingling and tremors. The driver said he just continued on his way.

Sometime later, Claude is at his place of work, 50 years, suffers from a herniated disc problem and cannot lift loads in excess of 30kgs. This morning, in the workshop, with his colleagues, he ranks a flange cargo.

They are light, he thinks, and he transports them to finger tip in pairs. Other colleagues stopped storing flanges and looked at him oddly: each flange weighs 34.5 kgs. Claude carries 138 kg without effort on fingertips... For two and a half months, the subject will be able, focusing few seconds, to lift weights up to 950 kg!

It will indicate that "image" forms in his head and then he understands that he can lift the masses. His colleagues mention that Claude seems "elsewhere" when performing these actions and itself does not always remember clearly what he has done in their presence.

By experimenting with a twist of an iron-bar of a diameter of 26 mm and by doing the test to a young colleague, they will see that Claude emits a force that will work onto his fellow and also twist the metal bar.

We believe that the phenomenon of object "rubber," is induced by the "demassification" partial, produced by bio-frequencies emitted by Claude.

No medical explanation is found to explain the "impossible" skills of the subject...

THE "MEMORIAL FLASHING" OF ABILITY AND KNOWLEDGE

In a rural area around Salers in France, 1994 Gerard C. and his family repeatedly observe spherical "flying saucers" near their home.

Soon after, although he has not studied botany, Gerard finds, without explaining, that he knows the name of many plants...

Better, he senses unconsciously exactly where the plant grows a few kilometers away...

Gerard also discovers he has therapist-magnetizer skills.

Gerard, his wife and their children will relate to "dream" or have strange sightings over several years, some similar in all respects with testimonials presented by Budd Hopkins, John Mack or Yvonne Smith.

OUT OF BODY EXPERIMENT (OBE)

The first difficulty is that different phenomena are designated with the same term "hotchpotch" of OBE (OUT OF BODY EXPERIMENT):

- endogenous phenomena produced solely by the brain
- exogenous phenomena with Meta Brain or soul-BUAWA where interfere...
- or phenomena related to the Aura

According to friend Christopher Blake, himself confronted with unsettling experiences, many cases of "abductions" are unrelated to the exocivilizations.

The person lives an experience for which she has no personal references or knowledge to analyze the phenomenon and put it in a safe environment, and stop it.

The experiment is reported to the top of falling asleep (*Paralysis hypnagogic*) or upon awakening (*Paralysis hypnopompic*), it is therefore present in waking and sleeping times.

The person finds himself paralyzed, unable to perform voluntary movements, and at the same time conscious of being in this situation.

To live this, there is no need to present clinical disorders, they are quite common conditions, but the intensity of which varies among individuals.

These states are common because they are related to the development of the nervous system and brain, it is a phenomenon of cerebral and glandular physiology.

In this particular state of being between waking and sleep, the person will perceive his room differently, and hear the sounds, smell the breath, sensations of elevation, vibration, electrical tingling, lights, etc.

Most troubling are the sensations when one feels that the hedges are moving, that the bed can move, a presence perceived as a threat, sensations of touches on the body and/or in the body, strong pressure, as one lies down on you, wants to come in you, or even attempts or experiences sexual intercourse.

All the senses are stimulated. Vivacity and duration distinguish it from a typical dream and form an experience that queries.

According to the scenario, the experience can go from panic until the state of ecstasy or pleasure.

It is very difficult for a person not to be panicked by this experience very real, and thus necessarily "traumatic". Consultation with a therapist will therefore present the diagnosis of a traumatic experience.

We find all ranges of observations in altered states of consciousness, like the famous experiments of "out of body" OBE (OUT OF BODY EXPERIMENT). These experiences are also at the time of falling asleep, or waking, or during sleep what is called "lucid dreaming".

These phenomena have been known for hundreds of years in different spiritual traditions and particularly in practices of yoga of dreams and in Clear Light of Tibetan Buddhism, and Dzogchen Bön.

Some practitioners look for them either as a "game" or as an essential practice in the Awakening routes or internal alchemy, which requires working on levels of consciousness.

Dream combines endocrines' functions, thyroid and nerve-related brain rhythms. According to Dr Lefebure the "out of body" experiences are due to brain physiology, special exercises allowing inter-hemispheric synchronization, some of which act on the centers of awakening and sleep, allowing to experience this dual status "know" that we dream, act on it, and also live the "out of body".

Misnomer term because in fact, it is an exploration of levels of consciousness which includes our reality. So you are not "out" but "in".

The difficulty of the analysis lies in the extreme embarrassment that we have to determine whether these cognitive processes are purely and strictly endogenous, produced only by the brain and endocrine functions, or if Meta-brain data or the soul-BUAWA interfere...

About me, I take into consideration the term OBE (OUT OF BODY EXPERIMENT) only to designate the phenomenon when Aura is disconnecting from the body, especially during a NDE.

Conclusion

The "memorial flashing" seems to be used by several exocivilizations for psychic experiences which unfortunately maintain a psychosis and a sense of understandable exocivilizations xenophobia.

These experimental practices are reprehensible and dishonest. We must remember that it is only psychological manipulations with false memories experimental purposes without malice.

The main problem is that it is a scientific program of terrestrials' abductions through secret agreements, illegal and illegitimate, with the armed wing of the oligarchic system, the black programs of the US Army. Ultimately, terrestrials "sold" represent approximately 95% of cases of abduction.

There is no doubt, some exocivilizations have capabilities, to perform technically or mentally, extremely powerful psychic manipulations.

However, according to my personal opinion, I think these "memorial flashings" of exogenous origin are a minority in relation to the numerous mental alterations of terrestrial origin, as are, for example, that lucid dreams are very common...

Thus, analysis of the testimonies of people who describe an abduction phenomenon must be made by taking into account natural endogenous psychic mechanisms and false memories. This is a great change in the overall perception of this...

*

123456789 10 **11** 12 13

EMERGENCE AND EVOLUTION OF MAN

Given the assumptions on the emergence of living, we must necessarily reconsider our view of the emergence of man.

There is a man, when the brain of a hominid was able to connect to its buawa, a kind of outer container psychic profile, able to capture the "moral laws" uaa issued by the entity woa and implemented by the BB-planetary.

The oemii that is to say, the man took in his only physiological dimension is associated with its buawa and becomes a oemmii involved in the great cybernetic loop of the evolution of the cosmos.

THE EMERGENCE OF HOMO HABILIS

We have a clear indication of the emergence of the first OEMMII that led to Homo sapiens, that is Homo habilis. Before, there are no humans, but only "simple" hominids.

Failing analysis technical means for identifying the presence of a krypton complex BAYIODUU in the brain of Homo habilis, we can just assume that its connection to BUAWA and his responsiveness to cosmic laws UAA induced a deist behavior.

From about 1 million years to 300,000 years Homo erectus, the likely descendant of Homo habilis, leaves traces of events of probable ritual cannibalism which are usually related to theology.

For Homo sapiens neanderthalensis, his nature of OEMMII is well identified with his funeral rites.

You can see on the chart. If on a branch of oyaagaa protomammals drifted in successive branches of mammals, if one of these phyla turned into primates, though they drifted various hominids until you get to homo habilis and subsequent branches, it was because of selection mechanisms and patterns of BB accelerated transformation in this derivation genotype.

Sooner or later the other animals would eventually be transformed into beings very similar to homo sapiens.

In other words: if the oemii of earth disappear, together with the apes, the monkeys, the platyrrhiniens, and even the rest of the mammals, the remaining classes would eventually crystallize (through an earlier and accelerated branching) in a new oemii.

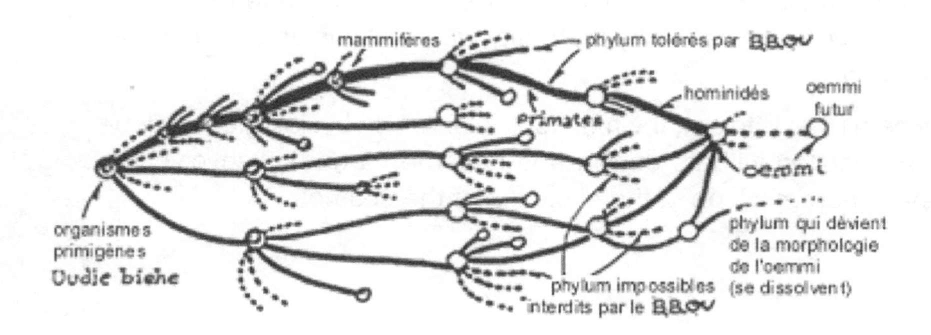

Since 2003, the majority of scientists consider two separate species: Homo neanderthalensis and Homo sapiens.

Nevertheless, took place in Europe, according to a 2010 study by the Neanderthal Genome Project, a part interbreeding between sapiens and neanderthalensis, there are 50,000—,100,000 years in the Middle East. So, this transferred to sapiens part of 1 to 4% to the genome of current Europeans.

Between the last 100,000 years and the last 10,000 years, the proto-historical knowledge of the evolution of Homo sapiens is very fragmented and unclear.

Therefore, many assumptions have been made about a possible interference exocivilizations on the evolution of Homo sapiens.

We saw in "Presence 1—UFOs, Crop Circle and Exocivilizations" elements of cosmic ethics that we called the "Pax Galactica" which exclude a socio-cultural interference not vital or inadequacy with the receiving civilization.

More importantly, the cosmological structure of interconnections between planetary-BB makes cosmobiophysically interdependent all human-OEMMII the cosmos.

Then, the link between humanity and its planetary-BB banned definitively the possibility of interference on the genetic structure of another humanity, for hybridization purposes, for example...

However, nothing prevents that very punctual and localized cultural exchanges have taken place in the past between exocivilizations and earthly peoples, if they were able to receive some information appropriate to their state of development.

It is heartbreaking to see how the belief is strong that the ancient Homo sapiens were brutes wearing skin beasts. It is likely that many ancient cultures have been more mature in some aspects of their development in comparison with the cultures of the world of today...

Sumerian mythology gave rise to numerous writings evoking ANUNNAKI, the Reptilians, the planet Nibiru, the Pleiadians. Some writing evoke these themes seriously respecting the original Sumerian writing, other are fantasies of a few liars ...

he popularization of these themes also made subjects of inspiration for Science Fiction...

This creates great confusion in the general public who do not really know what comes under serious written and argued on scientific with verifiable historical basis, or which is only mystification and deceit .... all this for the greatest happiness of Special Services responsible for misinformation... Also, will we clarify some points of this nebula thematic...

What is actually shown today in Sumerian writing?

What assumptions can be seriously developed?

COULD THE ANUNNA POSSIBLY BE ALIENS?

Sumerian mythology has given rise to numerous writings referring to the ANUNNAKI, the Reptilians, Planet Nibiru, and the Pleiadians. Some writing call forth these themes seriously carefully respecting the original Sumerian writing some whereas some other writings display a few fantasies... The popularization of these themes is also a subject of inspiration for literature Fantasy or Science Fiction... This creates considerable confusion among the general public who does not really know what is serious, what remains argumentative scientific writings and verifiable historical data, or what is tricky and fake ... and all this for the greatest happiness of Intelligence Services responsible for misinformation ... or disinformation. So the goal of this article is to clarify this issue that has become quite nebulous... What is actually demonstrated today in the Sumerians writing? What assumptions can be seriously developed?

Sumerian mythology describes the gods who are named ANUNNA creatures from the sky. The term Anunna is the generic word for the gods. The name of ANUNNAKI means Anunna settled on Earth (KI). We find these two words in the Sumerian texts; they refer to a real Sumerian myth... It is attested through historical documents, tablets with cuneiform writing.

Many authors will assume that the Gods may ultimately be alien beings. These alien beings whose origin no one so far has been able to identify the source through historical documents, or through the snippets of information we have about exocivilizations...

The analysis errors made by many authors, hoaxes and misinformation from official Intelligence Services, led to build a new contemporary UFO mythology... But so far, does this exclude the possibility that the Sumerians and other ancient populations would have been in contact with some exocivilizations? Shouldn't we try to avoid an overly simplistic approach and the risk to "throw away the baby with the bath water"?

HAVE THE ANUNNA CREATED MAN ON EARTH?

The Sumerian gods "ANUNNAKI" had planned to transform the human beings present on Earth before their arrival in order to transform them into slaves to their service.

An American writer named Zecharia Sitchin, launched into a long series of pseudo-scientific making through various books and made this type of human described in the Sumerian texts, transformed as a monkey, he claimed, into a human slave by ANUNNAKI. Kharsag texts, carefully translated by Anton Parks and published in his book Eden, show that this wild human was a man, not a monkey. For the transformation of this wild man into a docile slave there are multiple opportunities to explore...

Like many authors, Zecharia Sitchin unscrupulously asserted in a very categorical way that the ANUNNA would have genetically manipulated humans to alienate them. Some enlightened venal guys created juicy "religions" to the delight of institutional debunkers... Assuming that element of Sumerian mythology reflect this historical truth, an exocivilization has many ways to transform primitive human populations, without having to modify them genetically. An intervention on the human genome from an exocivilization is a simple hypothesis to interpret the Sumerian notion of "transformation" of the primitive human population. Moreover, such an eventuality in its reasonable assertion would be regarded as a small epigenetic change quite reversible after a few generations, or even as some very minor gene modifications.

Would an experienced exocivilization consider it has the right to intervene in the genome of any indigenous human species? This assumption, under both a eugenic and interventionist vision seems at least quite odd with the necessary intellectual maturity expected from an exocivilization able to achieve interstellar travel and fully aware of the interdependence of all species in the human cosmos.

Furthermore, this hypothesis is also inconsistent with the Oomomen's documents that attest ethics of human exocivilizations and rules overseen by the most advanced humans of the cosmos. According Oomomen, the DOOKAiens would observe our planet and actively participate in a comprehensive system of observation and protection of the Earth. This includes multiple species; it would serve the role of coordination and arbitration, for the scientific missions of different aliens that came to watch us.

"There is a race of OEMMII whose technology is beyond our understanding and that watches over different planets by scanning the OUEWA (interplanetary ships) which make incursions." (NR13, 14/04/2003—cf. The chapter PAX GALACTICA in "Presence: UFOs, Crop Circles and Exocivilizations")

In conclusion, the common sense of a terrestrial human, even if we were exiled on a planet in the company of natives that we would submit intelligently, it is likely that we got them under control with psychotronic devices ... already available in all good rays of Intelligence Services... This is a much simpler and much more effective than the genetic manipulation of an unknown species, whose effects may be substantially unpredictable...

In the Sumerian mythology, the project of turning humans to slave aborts through the intervention of a character called ENKI, which gives people the means to regain their freedom by educating them and providing them a big secret. This secrecy seems to have been the knowledge of metallurgy (see "The Awakening of the Phoenix" and "Eden" by Anton Parks). Enki will be assisted by several women of his clan, as his mother and Ninmah Mamitu-Nammu, the matriarch of the City of Kharsag. No text says that Enki is a Anunna god. Enki was a separate being, like his creator AN.

In the spirit of the Sumerians, the planets were gods ANUNNA "Mountains of Heaven." The Duku (DU-Ku) "holy mound" or "Holy Mountain" is the origin of Anunna. It is on this planet that the gods were created by AN (the father of Enki) and start the great rebellion of youth ANUNNA against the old gods represented by Tiamat, Queen Supreme. Enki led a clan called Nungal (or Igigi Akkadian) was in conflict with the Anunna.

These young gods ANUNNA themselves have genesis at least complex. About Enki, its nature is mysterious, even in the Sumerian tablets... The Nungal were the Watchers of the Bible and according to Anton Parks would be the servant of Osiris and Horus in Egypt. The following of Osiris and Horus are named in Egyptian Shemsu (from the Egyptian word SHMS follow, accompany). They were divided into several groups. In the apocryphal texts they would be giants. The Egyptians Shemsu would be near 2.10 m, while the size of normal humans ranged between 1.50 m and 1.65 m. They were there to protect the king Osiris and Horus. Anton Parks showed in his works that Enki was

Osiris in Egypt. The size differences can highlight ethnic groups or races. Neanderthal man, for example, had a significantly larger average size (about 1.85 m) that the average size of Homo sapiens which was contemporary (about 1.60 m). Similarly, the man of Flores had an average size very small, about one meter. This makes it very plausible hypothesis that Shemsu were an ethnic group specific. The Egyptian Shemsu they would have been a clan extraterrestrial called Nungal? This is a hypothesis that should be studied among others...

THE ANUNNAKI AND NIBIRU

To avenge their defeat, the ANUNNAKI pro-slavery banished ENKI who fled. As explained by Anton Parks in his works, no tablets are said to Enki had fled on a planet called Nibiru. Or even a planet named Nibiru would be the Anunna's one. This is a pure invention of Zecharia Sitchin found among others in his book "*The Lost Book of Enki*" and where he talks about a series of tablets that do not exist...

Zecharia Sitchin lied to his readers with this story and invented these "tablets" to provide false evidence of the existence of Nibiru—thesis he scaffolded his first book "*The Twelfth Planet*". "*The Lost Book of Enki*" is a monumental scam. Since a lot of media, magazines have taken this idea of "Nibiru—Planet of Anunna". As stated Anton Parks, the only place mentioned in the heavenly tablets as that of Anunna is the Duku and Enki in Sumerian mythology does not flee Nibiru, but in Africa, especially in Egypt. In Sumer, the secret temple of Enki and aquatic Abzu is called while in Egypt the temple of Osiris was water in the holy city Abdju (Abydos). Anton Parks was able to make a chronology of Nungal-Shemsu in his book "The Rise of the Phoenix".

There is obviously no connection either, between the false planet Nibiru and true planet Eris, which is beyond Pluto. But the irony of chance is also in the documents Oomomen. Indeed, as explained in the video documentary "*Presence and the Ummo file*" planet Eris discovered in 2003 by our astronomers, was reported 25 years ago, beyond Pluto, in 1979, in the documents alleged to exocivilization from the planet Ummo... Simple chance? Pure speculation? The planet Eris has a very atypical eclectic that made its discovery very difficult. These same Oomomen's documents give, 25 years before the discovery of the existence of this planet, the orbital position of this unknown planet... A simple probabilistic assessment makes it totally impossible for this level of predictability ... a quarter-century before the discovery the planet Eris. (*reference book PRESENCE 2—The language and the mystery of the UMMO planet disclosed*).

THE ANUNNAKI AND THE PLEIADES

The Pleiades and also supposed Pleiadians are derivative products and cream pie that reflect a deep disinformation, as always mixing truth and falsehood.

To a human eye, the Pleiades are seven stars. For an astronomer, it is a group of several thousands of stars... This group of some three thousand suns located about 450 light years of stars most likely include solar systems with exocilivilizations. This is highly likely. However, a semantic point of view, talk of "Pleiadians" has absolutely no sense... About which exocivilization in this constellation are we speaking?

References to the Pleiadians are to be taken with great caution... If one refers to the information contained in documents Oomomen, we can see that the most distant exocivilizations who visit us, are not located more than 150 light-years. With the usual means of our visitors and cosmological constraints that requires very long trips from already some 10 years. A journey of 450 AL supposes then a journey of about 30 Earth years. However, even if such a trip is possible, such a time, regardless of the longevity of human beings involved, makes it very difficult even for our friends from outer space... Among these multiple exocivilizations the Pleiades, it is very likely to have a humanoid phenotype, it is less likely that this phenotype could be confused those of Homo sapiens on earth. Simple matter of probability...

Sumerian tablets, no tablets are known explicitly states the Pleiades. Nevertheless, several tablets show the gods Anunna associated with the 7 stars visible that are supposed to be the Pleiades. For Anton Parks, Duku was in the Pleiades. Later, when the gods settled on Earth, they gave their city built in the mountains of Taurus name Duku (or Dukug) in tribute to their place of origin (cf. tablets translated in the book Eden). The Anunna and Nungal would have ended up on Earth, resulting in the battle against their queen Tiamat (see text of the

Enuma Elish). The ANUNNAKI are a group of expatriate warriors on Earth because of the war. At first, they had nothing else with them as the basic material that was in their "flying trolley". Exiles with means so rudimentary, they must use the local workforce...

The Anunnaki and reptilians

In the mythology of UFO disinformation, the aliens Anunna would have a human form, and the Illuminati would be their descendants.

Following the Ummo documents we can understand that our planet is in the hands of three major human oligarchic groups (Western, Russian and Chinese) "dangerous, cynical and corrupt." ref. D1378

The hypothesis of extraterrestrial origins of these oligarchic groups seems to me wacky. I feel that, once again, disinformation is at work and she tries to hide these dangerous oligarchic groups truly terrestrials, behind a smokescreen of preposterous Illuminati aliens... This gives another good reason to give the full powers to these dangerous oligarchic groups that pretend to want to "protect" the humble citizens of the supposedly wicked Illuminati aliens...

A world dictatorship willed, would be the apotheosis of a maneuver in the pure precepts of Machiavelli... Machiavelli teaches that *the Prince should be feared, but still not to be hated. If he is hated, he returned to the people against him, only if he is feared he maintains its authority and power. So it is from this point of view of good practice is to maintain fear, without it turns into hatred. People kept in fear remains "calm". He dares not stand up against the power. People begin to hate his sovereign will seek to overthrow him and he will follow those who lead the revolt. All tyrants of humanity know that. There is a calculated skill, cunning, Machiavellian to handle insecurity and use fear.*

As for the reptilians who also embody disturbing characters, Zecharia Sitchin had quarreled about it with David Icke. Zecharia Sitchin has knowingly ignored the various Sumerian records where one sees many gods with a form of "crocodile" or by extrapolation "reptilian". Zecharia Sitchin wanted that is thesis glue up with the Bible. To him the gods could not be in the form of the Biblical Serpent! But seeing that the "tide reptilian" was gaining momentum, he did not insist and finally avoided the subject just to surf this wave.

Several hypotheses can be advanced to explain why the gods Anunna are sometimes represented by a form of "crocodile".

The first assumption is that the references to animal kind of reptilian and possibly their connection to the human race are the memories of a distant memory collective conveyed through the ages. (See book by Anton Parks EDEN)

The texts of the first Bible themselves seem clearly to rely on written Sumerians. It is likely that they themselves were the first vectors of a written transposition of a long oral tradition. The memory of traditional knowledge seems to have conveyed through the ages that reptiles to mammals predated. These ancestral knowledge, broadly confirm the modern science of the twentieth century and into the brain with the first evolution theory in the sense of the tri-unique brain of Paul MacLean in 1969. This highlights, in passing, that our ancestors, far from being idiots in animal skin armed with a club, where on the contrary, very fine in their world observations and their analyzes... If this had not been the case, humanity would anyway not even reached our stage of development...

It may nevertheless oppose an objection to this first hypothesis, to formulate the second hypothesis. The racial dissymmetry highlights different ethnic groups. Neanderthal man, for example, had a significantly larger average size—about 1.85 m—that the average size of Homo sapiens which was contemporary—about 1.60 m. Similarly, the man of Flores had an average size very small, about one meter. This then makes a very plausible assumption that the Egyptians Shemsu was a specific ethnic group.

The Sumerians refer to "the jaws ruthless giant snakes." Did they find dinosaur fossils sufficiently evocative of the past of these carnivorous animals? Or could they have other sources of information? Then we can ask ourselves why and how these "furious dragons" could be transformed to become "like the gods"? The Egyptian Shemsu would they have been a clan of extraterrestrials called Nungal? This is a hypothesis that should be studied among others...

New hypotheses about the Sumerian mythology

The theories of Zecharia Sitchin have reached a great popularity because the author started by doing a good quality scientific work. His research was based on the analysis of Hebrew texts deriving from the Sumerians ones and interpretations of Sumerian cuneiform documents. Afterwards though, our researcher kept on traducing Hebrew words and Sumerians texts without mastering these languages enough.

Indeed, despite his apparent good will, Zecharia Sitchin makes serious errors of translation, at the point of even suspecting that he knew how to translate the Sumerian... He never used, in any of his books, a cautionary formulation such as "here's my own translation of that text..." or similar; his translations seem to come out of nowhere, often without any reference. This is the case for translations either from Hebrew, that he did not master very well, or from Sumerian, where mistakes are even more damaging and preclude a real understanding of the texts... Zecharia Sitchin reached such a level of misinterpretation and speculations that he just ended in saying nothing more than a nice story ... but very far from the reality of the original texts. The rantings of Zecharia Sitchin led him to create a new contemporary UFO mythology that is certainly regarded as delightful by Intelligence Services in charge of disinformation...

But is all this enough to exclude the possibility that the Sumerians and other ancient peoples might have been in contact with exocivilizations? Don't we risk taking an overly simplistic approach and don't we risk to "throw away the baby with the bath water"? This is exactly what Intelligence Services would like us to do...

The Sumerians and the peoples of the Indus Valley in that period (approximately - 6,000 to - 5,000 years BC) had highly developed cultures potentially interest for visitors from outer space... What would be the elements that might reveal a potential contact with exocivilizations? Sumerian mythology says that the ANUNNAKI gods are "Crocodile-like" hominid beings in some seals (small Akkadian or Sumerian seals on clay, printed in pieces of clay) that the Sumerian gods had a reptilian appearance, but it is never clearly said in the texts. A few texts from Kharsag translated by Anton Parks in EDEN show Enki and Ninmah bearing reptilian names with some "crocodilian" physical descriptions together with some references where characters like Inanna-Ishtar and her lover Dumuzi are reported to have faces or eyes of a Umshumgal

(Great Dragon). By imaginative extension, the "crocodilian" hominid is called "Reptilian" in the mythology of David Icke and some others...

As we explained in the book "Presence: UFO, Crop circle and exo-civilizations" the Oomomen's documents mention the visit of humanoid beings that have a "scaly skin". This evokes obviously a physical "Reptilian" characteristic, but a scaly skin is in no way an indication of the class of a alive being. A humanoid with a "scaly skin" could well be a mammal ... or a fish!

In 2007, in *"PRESENCE UFOs, Crop Circles and Exocivilizations"* we presented 18 exocivilizations hereby permanently on Earth.

In 2015, they are now 23.

The most numerous, 19 exocivilizations are benevolent and act positively on the future of terrestrial humanity. Some 4 others may be considered "immoral" in their practices, such as the case of GOHOians.

Another disturbing element which could also support the hypothesis that the Sumerians might have had contact with this kind of visitors from outer space is the chronology mentioned in the Oomomen's documents. Indeed they indicate that ET humanoids called by them "2ians" have been visiting the Earth for 30,000 years. They are supposed to come from a star located 47 LY away, and have a great height, a scaly skin, which is also fully compatible with the chronology and the Sumerian texts' data.

The "2-ians" are "tall" is between 2 and 3 meters.

As indicated by Anton Parks, the Followers of Enki, the Egyptian Shemsus are about 2.10 m while the height of normal humans ranged between 1.50 m and 1.65 m.

The "2-ians" are known by our friends from UMMO, for their scaly and pale skin, the head of "snake" and their whistling speaking, which spread rumors of "reptilians".

It should not be overlooked, either, that the Sumerian tradition presents the "Serpent People" as beings of power and knowledge. They inhabit the high mountains of Kurdistan, from where they came to bring to men the benefits of civilization.

In conclusion, despite many stories about it which are totally spooky, and probably profit to misinform the seriousness of the information itself ... so do not exclude the real and serious possibility that Sumerians and

Akkadians may have been in contact with the "2-ians" that are always present on Earth.

Ashiwi Native Americans and the Aliens

Ethical contacts between terrestrials and exocivilizations could they leave traces into the indigenous cultures? Elements may suggest that Indian Ashiwi have also been in contact with exocivilization, about 3,000 years ago.

For the first presentation of my research in the USA, in February 2010, at the UFO Congress in Laughlin, I had an amazing visit: the Elder shaman of the Zuni tribe. Or rather Ashiwi as they call themselves, the official name of Zuni has been given by the Spanish. The Elder shaman Mahooty Clifford, came to see me as if he had known me forever, and immediately designated me as a "brother"!

This surprised me greatly, but our mutual sympathy has not diminished over the course of our meetings, on the contrary, the convergence of views and our exchanges were of the richest...

The territory of the Ashiwi (which means "those where everyone is special"), is located southwest of the city of Albuquerque. This area is crossed by the Zuni Canyon leading to the Grand Canyon, farther west. Clifford Mahooty quickly informed me that the Zuni territory is "UFO-active" since the 60s. The UFO sightings are truly dazzling, 1 to 2 times per month, often arriving from the south of Zuni Canyon, sometimes during the day. On several occasions UFOs were low to the ground. As in 2001, when probably automatic vehicle of 3 meters in diameter arose a few dozen yards from his home in an area of the corn crop. Since, at this location the vegetation is stunted, and corn does not grow at all.

Several other people from the tribe, including a Zuni archaeologist, one evening in 2005 on a road near the village found themselves face to face with a craft more than fifteen meters in diameter hovering low to the ground. One of the frightened Indians wanted to shoot the machine with a hunting rifle large caliber, but my friend Dan, the Zuni archaeologist, dissuaded him... They stood at a distance until the vehicle leaves with a few oscillations, and lightning speed...

Elder chaman Clifford Mahooty and Denis Roger Denocla

UFO congress Laughlin Nevada USA 2010

Here are some examples of meetings that can be done in the territory of Zuni, and all residents are aware of the phenomenon. For much of them, this is "normal", especially as the Elder shaman of the tribe of Zuni will tell me the story and mythology of the tribe. This is markedly different from the official version, as it can be found on which pages of "wiki-censorship."

In this mythology, the origin of the culture of the people comes from the Zuni Kachinas. The Kachinas are represented by colorful characters who embody and masked during ritual dances, not just the "spirits" as the politically correct want to have us believe. But all the principles of cultural society Ashiwi, with an angle very special... Take the example of a conventional type Kachina "children of incest" are hideous and crazy characters that symbolize the dangerous effects of potential relationships of incest and inbreeding. Each family is responsible for the transmission of the oral history of a type of Kachina with all its symbolism. What we know is that for most of the Kachinas, the origin of each symbol is associated with …. a people of the Cosmos!

Another point totally ignored is that giant Kachinas are the cause of the foundation of the Zuni culture, his moral and social values.

Clearly, Mahooty Clifford explains that the Zuni have been in contact with an extraterrestrial civilization that would have given them a lot of knowledge. Clifford tells me especially in the field of astronomy, the Zuni have known for "always" the existence of certain nebulae completely unknown before the advent of powerful telescopes.

For me, the information about the Kachinas giants that would be the cause of the foundation of the Zuni culture, immediately evoke something specific...

Indeed, I draw the attention of Clifford on the framework of alien races that I published in the first book Presence. The aliens whom I have called "IOXiens" are about 3 meters high, which would have started their visits to Earth-896 before JC, a presence there initiated some 3,000 years... The assumption that the Giants are the Kachinas-IOXiens fit pretty well and seems very plausible Clifford Mahooty, both for the chronology and the moral content of this specie. From the source of the UMMO letters, the IOXiens people have a high intelligence and a strict moral. They are therefore good candidates to have issued guidance to entities, on the one hand, Zuni, Hopi and other. This former exocivilization would have intervened long before the "Pax Galactica" for 80 years, at a time when the level of acceptance was easily managed by visitors, so infrequent. This gives us a good hypothesis for this old exocivilization. But why such a "UFO-activity" since the 60s?

I ask the Elder shaman of which is for me the logic critical key points. Did he know rumors of secret nuclear facilities in the territory of Zuni? Such facilities in the sixties would not have failed to generate active surveillance by the exocivilizations we meet. Clifford does not exclude this hypothesis; the work could not be done without the knowledge of the tribe.

The Zuni territory included an inactive volcanic area. This type of geological activity uses to interest many exocivilizations and could be a source of curiosity and learning for our visitors. However, this volcanic cannot be correlated with the period of 60 years. But, I suggest to Clifford that the volcanic activity and the monumental canyons may have been one of the reasons that might motivate IOXiens to come in the region 3,000 years ago.

Although the assumption of civil and military nuclear facilities is frequently the cause of an area "UFO-active," I take into account a particular data, the history of the relationship between the Zuni and the probable IOXians. This relationship had been former require the establishment of a logistics IOXien's infrastructure. This underground base can always be active including other exocivilizations. Or, this could be an "exo-historical place", a kind of "Statue of Liberty" site for exocivilizations...

This hypothesis also holds the attention of Clifford, who tells me he knows also the existence of petroglyphs that have never been studied in this area difficult to access. The historical roots of the relationship between the Zuni and exocivilizations should not fail to be shown over time by petroglyphs. Another topic of surprise is in the language of Ashiwi whose origin is a complete mystery. This language has nothing in common with other Indians. Why?

Conclusion on the influence of exocivilizations

We must beware of hasty and simplistic conclusions, and beware of conventional ideas conveyed perniciously to confuse... And always ask: "Who's gonna get the benefits?"

Although Sumerian mythology had only tenuous historical basis, it has nonetheless been a founding mythology of the great religious texts of the Middle East.

This greatly undermines these religious dogmas and the simple assumption that this mythology may also have a close link with some exocivilizations could be a fatal blow to the anthropocentric dogma...

Possible future developments of man

The WOA generator appears to have had the good idea to generate infinity of cosmos of which follow the same patterns of structuring with variations, but always in fine with beings capable of playing his expected cybernetic role.

According Oommomen's calculations, there would be 10^{30} human races in the multi-cosmos WAAM-WAAM.

As for the forms of living beings, only a few millions of billions of forms would be possible on each planet.

Normally, a node of a phylum (tree) can give birth to about two hundred twenty thousand (on average) new branches or phyla through a directed mutation, that is to say controlled. In some nodes were detected (approximately) 18,376,000 possibility mutations tolerated by the planetary-BB.

Can we study the phylogeny as possible on the various OYAA (planets) of the WAAM-WAAM? Obviously not! It may be that the possible living beings are counted by trillion or quadrillion.

We calculated that only our cosmos WAAM could encode up to

5.2×10^{18} 1018 models, but the inaccuracy of the calculation is suspected that they could be much more. These primary patterns can be derived from thousands of millions (individuals or copies), so that the order of magnitude for all WAAM-WAAM would reach a figure of copies "possible" different $10^{,526}$ (order of magnitude).

The information contained in OOMMO documents clearly explains the outline of the evolution of living beings and man:

• First stage: An organism has simple reflexes, and responds directly to the stimulus of the physical environment.

• Second stage: An organism able to process information and directs his conduct not only deterministically, but under his stored information.

• Third stage: An organism (OEMII—human) whose brain has experienced a quantum jump which allows him to have a conscience, relatively free, and connected to BUAWAA (Psyche) at BUAWEE BIAEII and whose conduct contributes to shaping the WAAM-WAAM

• Fourth stage: the OIXIOOWOA. The probability that arises a mutation OIXIOOWOA (One in a brain determined) and over the first ten million years of a network of OEMMII, is very high (probability close to one) and that reaches the unit, if it has passed at least thirteen million years. It is very rare that over 15-20000000 years (if humanity survives) it produces an identical mutation. The frequency distribution law in time following a very singular function, then we expose graphics. In each social network from any planet, one individual of this species named OEMMIIWOA, is generated by chance for the first time.

Then it takes a wide range of time without the occurrence of the phenomenon, it will not happen until several million years.

The mutation OIXIOOWOA therefore generates a type of brain radically different from that of the OEMII which it proceeds.

The OEMIIWOA and consistent, is a new biological species with a distinct genome.

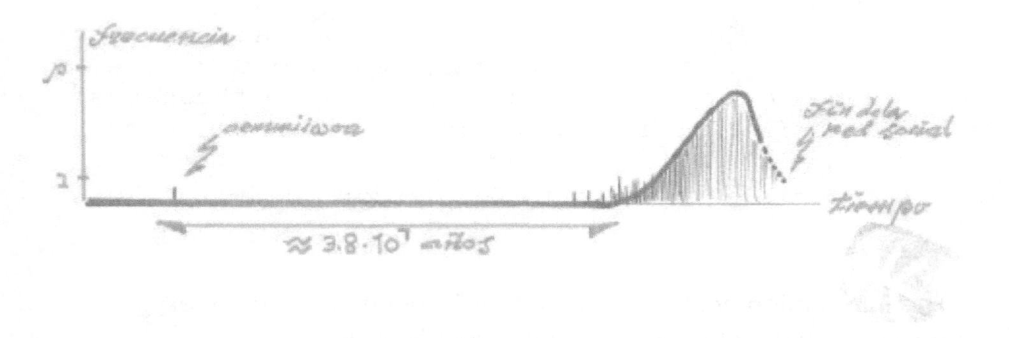

Past thirty-eight million years, it occurs a surprising biological-physical phenomenon. Almost all brains have mutated.

However six million years before that, hundreds of thousands of OEMII beings had experienced this mutation which converted them into OIXIOOWOA brain.

After a few years (no more than fifty years), the human body that houses the OIXIOOWOA brain "DISAPPEARS". (Note that we do not say that he may die but "disappears"). But it is clear that if the entire population that inhabits the planet reaches this stage, mankind ceases to exist.

We have seen through the previous books that many exocivilisations essentially benevolent were present on Earth, sometimes since tens of thousands of years. As part of the Earth's human evolution, every citizen of our planet can legitimately ask the questions:

How are organized our visitors on their planets?

How do they manage their planet?

What is their political organization?

What economic structure do they have?

So many questions that can make us think about our own socio-economic development. What about the demosophie, the societalism, the ethocratie?

What kind of society could we imagine for tomorrow?

This is the subject of the discussion we will see in PRESENCE 4 “Toward a New World ... with the Exocivilizations”.

12

GENERAL CONCLUSION

Through Theory 3 Third, I tried to present some answers to great traditional questions of humanity.

Where we come from, what is the place of man in the universe, what might our future?

Thus, these assumptions provide new explanations beyond the limits of our current knowledge.

Our Universe, at least decadimensionnal, would be made up of sheets of pairs of cosmos, driven by a global Meta-Brain. The planets themselves are connected and driven by their specific planetary Meta-Brain, through a LEIYO effect on krypton would happen on the threshold of the "krypton constant".

This would boot a process that will lead to the emergence of living in lifeless matter. Thus, the chain of krypton BAAYIODUU pilots amino acid groups to form RNA-archaic, which will quickly be encapsulated to create the first self-replicating living entities.

The great cycle of directed evolution is started under the control of the chain of krypton BAAYIODUU, it follows the laws of cosmobiological phylogeny and orthogenesis that will lead to OEMMII Human beings in the cosmos.

These are partly under the control of the autonomous informative entity conformation, the BUAWA psyche.

This new cosmological framework streamlines many subjects which fell within the "paranormal activities" or Metaphysics.

More than ever, science and conscience are found together to help move us to a world we all want more just, more peaceful and happier.

In a distant prospect, perhaps we will reach those wisest human civilizations which evolve into a species whose level of complexity will be equal to their cosmological driver, the planetary-BB...

Some of these assumptions could be verified experimentally today. The inspiration of these assumptions is due to documents being claimed by an exocivilization, this gives the whole process a character unusual and an extreme psychological difficulty for the reading and acceptance.

But such are the facts.

Thus, by placing my trust in history, the most daring bet I can do today, is that the assumptions of the Theory of 3 Third, just be studied and tested...

The author strictly prohibits reference to his research for religious purposes.

"When we violate a divine standard, we do so based on an entropic attitude. Every social sin, every sin against what you call charity (love) dissolves to a lesser degree the coordination of a social network.

If I cause harm to my brother, I can cause inhibition of its observer functions, I contribute to a certain level to slow the uptake of information WAAM plan, that is to say that I am helping to create the ENTROPIE, the DISORDER, slowing the progress of Pluriunivers.

This deserves condemnation from all OEMII of the WAAM-WAAM, since it significantly affects US."

MANIFESTO FOR THE EXOCIVILIZATION'S RECOGNITION

This manifesto outlines a few of the basic principles necessary to establish a fair and long-lasting relationship with any exocivilization respecting the "Pax Galactica":

Exocivilizations' Rights

- Official recognition of the exocivilizations
- Application of Human Rights to all exocivilizations
- Application of the Geneva Convention to all exocivilizations
- Restitution of the bodies of dead explorers to their exocivilizations (in reference to the 1994 Law on Bioethics)

Exocivilizations' Duties

- Respect for the UN conventions and resolutions
- Respect for the rights of States
- Respect for the rights and integrity of people and property.

D. R. DENOCLA

"Knowledge for whom? Knowledge for what?"

BIBLIOGRAPHY

The sources of OOMMO materials are from the site www.ummo-sciences.org and www.ummo-ciencias.org and D.R. DENOCLA.

Alexandre Oparin « L'origine De La Vie », 1938, éditions Masson

Andréï Sakharov « Œuvres (scientifiques) complètes » Edition Anthropos (ouvrage disparu des catalogues!)

Budd Hopkins "Intruders: The Incredible Visitations at Copley Woods" Three Rivers Press; Édition, 1992

Christian de Duve « A l'écoute du vivant », éditions Odile Jacob, 2002

Francis Crick, « La vie vient de l'espace », édition Hachette, 1981.

Daniel Verney « L'Astrologie et la science futur du psychisme », édition Le Rocher, 1987, Monaco.

Denis Roger Denocla « Acid Jones et le mystère du temple de la science » édition ADDOM, 1990.

Dr. Hyman "The Mischief-Making of Ideomotor Action" in the Fall-Winter 1999 issue of The Scientific Review of Altrnative Medicine, ©1999, Prometheus Books.

Jacques Pazelle, communications personnelles.

John Maynard Smith et Eörs Szathmary, « Les origines de la vie », éditions Dunod, 2000

Ludwig Von Bertalanffy : « Théorie du système général » Edition Dunod, 1993.

Marie-Christine Maurel « La Naissance de la vie », éditions Dunod, 2003

Michel Marcel, communications personnelles.

Percy Seymour "Astrology: the Evidence of Science", Arkana, édition Penguin, 1988, Londres.

Stephen Jay Gould "Ontogeny and phylogeny",1997, editions Belknap Press (janvier 1985)

Stephen Jay Gould, « La structure de la théorie de l'évolution », 2007, NRF - Gallimard

Tsiang Kan Zheng, revue AURA — Z n° 3, 1993

Yvonne Smith "Chosen. Recollections of UFO abductions through hypnotherapy", 2008, éditions Backstage Entertainment.

Webographie:

http://fr.wikipedia.org

http://plato-dialogues.org

http://www.antonparks.com

http://www.astrosurf.org

http://www.cafe.edu

http://www.cropsciences.org

http://www.futura-sciences.com

http://www.gillescosson.com

http://www.mineralinfo.org

http://www.pnas.org

http://www.quackwatch.org

http://www.scedu.umontreal.ca/profs

http://www.sciencedirect.com

http://www.societechimiquedefrance.fr

http://www.quanthomme.free.fr

http://www.morpheus.fr

8 Esp. de la Manufacture
92136 Issy-les-Moulineaux

Imprimé par :
Graphic Systems.Com
69 chemin de la Chapelle St Antoine
95300 Ennery

Achevé d'imprimer en septembre 2013
Dépôt légal : septembre 2013
Imprimé en France

Made in United States
Cleveland, OH
05 June 2025

17529863R00136